Sustainable Combustion Systems and Their Impact

Sustainable Combustion Systems and Their Impact

Editors

S.M. Ashrafur Rahman
Islam Md Rizwanul Fattah

MDPI • Basel • Beijing • Wuhan • Barcelona • Belgrade • Manchester • Tokyo • Cluj • Tianjin

Editors
S.M. Ashrafur Rahman
School of Mech., Medical &
Process Engineering
Queensland University
of Technology
Brisbane
Australia

Islam Md Rizwanul Fattah
School of Information, Systems
and Modelling
University of Technology
Sydney
Australia

Editorial Office
MDPI
St. Alban-Anlage 66
4052 Basel, Switzerland

This is a reprint of articles from the Special Issue published online in the open access journal *Energies* (ISSN 1996-1073) (available at: www.mdpi.com/journal/energies/special_issues/ Sustainable_Combustion_Systems).

For citation purposes, cite each article independently as indicated on the article page online and as indicated below:

LastName, A.A.; LastName, B.B.; LastName, C.C. Article Title. *Journal Name* **Year**, *Volume Number*, Page Range.

ISBN 978-3-0365-1396-6 (Hbk)
ISBN 978-3-0365-1395-9 (PDF)

Contents

About the Editors

S.M. Ashrafur Rahman

Dr S M Ashrafur Rahman is currently working as an Environmental Consultant at Trinity Consultants Australia. He is also working as a Visiting Fellow at Queensland University of Technology (QUT). He obtained his Doctoral degree in Mechanical Engineering from QUT in 2018. Previously, he completed his Master of Engineering Science from the University of Malaya in 2015 and Bachelor of Science in Mechanical Engineering from Bangladesh University of Engineering and Technology in 2012. His research work focuses on evaluating various bio-oils as possible supplements to petro-diesel.

Islam Md Rizwanul Fattah

Dr IMR Fattah is a Postdoctoral Research Fellow in the School of ISM, Faculty of Engineering and IT at the University of Technology Sydney (UTS), researching the effective use of waste for sustainable energy applications. He has accomplished his Ph.D. in reducing PM and soot emissions from diesel combustion from the University of New South Wales (UNSW, Sydney) in 2019. Previously, he completed his Master of Engineering Science from the University of Malaya (UM) in 2014 and Bachelor of Science in Mechanical Engineering from Bangladesh University of Engineering and Technology (BUET) in 2011. He has been actively engaged in the field since 2012 by publishing over 60 articles and gaining over 4500 citations of his works. He serves as an Editorial board member at *Energies* (MDPI). He also has peer-reviewed over 200 journal articles of many WOS-indexed journals throughout his career.

Article

Maximising Yield and Engine Efficiency Using Optimised Waste Cooking Oil Biodiesel

Luqman Razzaq [1], Shahid Imran [1], Zahid Anwar [1], Muhammad Farooq [1],
Muhammad Mujtaba Abbas [1,*], Haris Mehmood Khan [2], Tahir Asif [1], Muhammad Amjad [1],
Manzoore Elahi M. Soudagar [3], Nabeel Shaukat [2], I. M. Rizwanul Fattah [4,*]
and S. M. Ashrafur Rahman [5,*]

[1] Department of Mechanical, Mechatronics and Manufacturing Engineering, New Campus,
University of Engineering and Technology, Lahore 54890, Pakistan; luqmanrazzaq@uet.edu.pk (L.R.);
s.imran@uet.edu.pk (S.I.); zahidanwar@uet.edu.pk (Z.A.); engr.farooq@uet.edu.pk (M.F.);
tahir.asif@uet.edu.pk (T.A.); amjad9002@uet.edu.pk (M.A.)
[2] Department of Chemical, Polymer and Composite Materials Engineering, New Campus,
University of Engineering and Technology, Lahore 54890, Pakistan; hariskhan@uet.edu.pk (H.M.K.);
2015ch284@student.uet.edu.pk (N.S.)
[3] Department of Mechanical Engineering, Faculty of Engineering, University of Malaya,
Kuala Lumpur 50603, Malaysia; me.soudagar@gmail.com
[4] School of Information, Systems and Modelling, Faculty of Engineering and IT, University of Technology
Sydney, Sydney, NSW 2007, Australia
[5] Biofuel Engine Research Facility, Queensland University of Technology, Brisbane City, QLD 4000, Australia
* Correspondence: m.mujtaba@uet.edu.pk (M.M.A.); rizwanul.buet@gmail.com (I.M.R.F.);
s2.rahman@qut.edu.au (S.M.A.R.)

Received: 6 October 2020; Accepted: 11 November 2020; Published: 13 November 2020

Abstract: In this study, waste cooking oil (WCO) was used as a feedstock for biodiesel production, where the pretreatment of WCO was performed using mineral acids to reduce the acid value. The response surface methodology (RSM) was used to create an interaction for different operating parameters that affect biodiesel yield. The optimised biodiesel yield was 93% at a reaction temperature of 57.50 °C, catalyst concentration 0.25 *w/w*, methanol to oil ratio 8.50:1, reaction stirring speed 600 rpm, and a reaction time of 3 h. Physicochemical properties, including lower heating value, density, viscosity, cloud point, and flash point of biodiesel blends, were determined using American Society for Testing and Materials (ASTM) standards. Biodiesel blends B10, B20, B30, B40, and B50 were tested on a compression ignition engine. Engine performance parameters, including brake torque (BT), brake power (BP), brake thermal efficiency (BTE), and brake specific fuel consumption (BSFC) were determined using biodiesel blends and compared to that of high-speed diesel. The average BT reduction for biodiesel blends compared to HSD at 3000 rpm were found to be 1.45%, 2%, 2.2%, 3.09%, and 3.5% for B10, B20, B30, B40, and B50, respectively. The average increase in BSFC for biodiesel blends compared to HSD at 3500 rpm were found to be 1.61%, 5.73%, 8.8%, 12.76%, and 18% for B10, B20, B30, B40, and B50, respectively.

Keywords: biodiesel; waste cooking oil; transesterification; response surface methodology; central composite design

---

1. Introduction

Over the last few decades, the rapid decline of fossil fuels has become a significant problem. On the other hand, the energy demand is continuously rising owing to the rapidly expanding population coupled with the increased rate of urbanisation [1–3]. This scenario demands the adoption of alternative energy resources to address the issues related to the energy as well as the environment [4–7]. Recently, biofuels produced from a variety of naturally occurring resources [8,9] have emerged as an alternative energy source owing to their comparable physicochemical, performance, and emission characteristics [10–12]. The main hindrance in the commercialisation of biodiesel is its production cost, and the production cost mainly depends upon the feedstock oil used. The use of edible oils for the production of biodiesel may cause an imbalance in the food chain, and has sparked the 'food vs. fuel' debate for decades. Waste cooking oil (WCO) can be a potential alternative to edible oil for biodiesel production. WCO, when disposed of without treatment, causes environmental pollution. The use of WCO for biodiesel production not only reduces the production cost of biodiesel, but also decreases the environmental burden.

Generally, biodiesel can be used as a blend (with or without additives) with diesel in diesel engines without any modification in the engines [13,14]. Biodiesel can be converted to fatty acid esters by reacting fatty acids and short-chain alcohol through a process known as transesterification [15,16]. Homogeneous alkaline catalysts such as KOH, NaOH, and CH_3ONa have been used as catalysts in this process [17,18]. Methanol is suitable for the transesterification process [19]. The biodiesel yield depends upon the operating parameters of the transesterification process, which include temperature, catalyst concentration, methanol to oil ratio, reaction speed, and reaction time [20]. The operating parameters should be optimised to obtain the optimum yield of fatty acid methyl ester (FAME) [21]. The effectiveness of the transesterification process is assessed by the reaction kinetics, mass transfer, and equilibrium in the reaction mixture. The results of this process are used to predict the conversion yield and design a model which can be used for the prediction of conversion yield [22].

Performance of CI engines operated with biodiesel depends upon different factors including, but not limited to, compression ratio, injection timing and injection pressure [23,24]. A slight reduction in brake thermal power and brake torque was observed when biodiesel tested on the six cylinder DI diesel engine [25]. Biodiesel blends significantly reduced emissions such as carbon monoxide, carbon dioxide with a slight increase in oxides of nitrogen NOx [26,27]. A previous study by Nirmala et al. [28] tested pure WCO based biodiesel (WCOBD) and compared the results with that of pure diesel. WCOBD showed 4.2% higher BSFC, 3.6% lower BTE, and 10.8% lower BP than conventional diesel. Akcay et al. [29] studied hydrogen addition to intake air in conjunction with 25% WCO biodiesel blend (B25). The effects of hydrogen on the BSFC of B25 fuel were not significant compared to diesel fuel. Can [30] studied the engine performance of 5% and 10% blend of WCO with diesel and reported a slight increase in BSFC (up to 4%) and a small reduction of BTE (up to 2.8%) with the addition of the biodiesel for all tested engine loads.

Many techniques have been reported to optimise biodiesel yield. RSM is one of the most widely employed techniques [31–34]. RSM is a mathematical technique used for empirical model building. Yield is known as the response which depends upon independent variables which are the operating parameters of the transesterification process. RSM develops a suitable experimental design model to provide optimum operating conditions [35]. Jamshaid et al. [36] used the RSM technique for the optimisation of biodiesel yield from cottonseed oil and reported an optimum biodiesel yield of 98.3%. Mostafaei et al. [32] examined the effect on the biodiesel yield of the independent variables reactor diameter, ultrasound strength, and liquid height. The RSM technique was used to develop interaction among these independent variables. Anwar et al. [37] investigated the yield optimisation of second-generation biodiesel produced from Australian native stone fruit oil using the RSM technique, and thereby an optimum yield of 95.8% was obtained. This article focuses on optimising the biodiesel production process from WCO of Pakistani origin using the RSM technique that involved CCD [38].

The prevalent energy crisis has adversely affected the global economy. The economies of many developing countries, like Pakistan, have gone uncompetitive due to the shortage of usable energy. With large agriculture land and a population of over two hundred and twenty million, Pakistan is one of those economies which has the potential to generate large quantities of renewable power. Besides, waste recovery after the use of edible oils is almost non-existent in such economies. This makes the problem more complicated. If properly treated, WCO has the potential to partially substitute some of the non-renewable fuels used for energy generation in the transportation, domestic, or industrial sectors. This present study is one such effort that investigates the possibility of using WCO in a diesel engine.

2. Materials and Methods

2.1. Materials

Work reported in this study is based on WCO collected from four different restaurants. Any suspended particles in WCO were first removed using filter papers of size 12.5 mm diameter. The filtered WCO was heated at 100 °C for 1 h to remove the moisture content, followed by the cooling process. HSD for preparing biodiesel blends was purchased from local market. Methanol, ethanol, sulphuric acid, potassium hydroxide, and phenolphthalein were purchased from Sigma Aldrich (purity > 99%).

2.2. Biodiesel Production Using WCO

The physicochemical properties of WCO were illustrated in Table 1. Density and viscosity of WCO were observed higher than original canola oil (the source oil) due to the formation of unsaturated bonds through continuous use. This may also result in a high acid value of WCO. The acid value (AV) should be decreased before the conversion of WCO into biodiesel using mineral acids. Therefore, the raw WCO was treated with mineral acids (H_2SO_4, HCl, and H_3PO_4) through a process known as esterification for reducing its free fatty acid (FFA) contents, which in turn dictates the AV [39,40].

Table 1. Physicochemical properties of WCO.

Properties	Units	Values
Density at 15 °C	kg/m^3	910.30
Acid Value	mg KOH/g	7.80
Free fatty acid	%	3.90
Molecular weight	g/mol	860.56
Viscosity at 40 °C	mm^2/s	6.80

The most important factor in the esterification process was the quantity of methanol. Increasing the methanol concentration in the mix would result in more effective FFA reduction. Other parameters were, reaction speed that was 600 rpm, the temperature was 60 °C and time for this reaction was 3 h. In this process amount of methanol used was 2.25 × FFA and the amount of sulphuric acid used was 0.05 × FFA. In this study, the maximum reduction (74.7%) in FFA was observed by treating WCO with H_2SO_4, which was followed by H_3PO_4 (63.1%) and HCl (54.9%).To reduce the FFA of the WCO, the oil was treated into two steps using H_2SO_4. In the first step, AV was reduced from 3.9 to 1.45 mg KOH/g and in the second step it was reduced to 0.34 mg KOH/g. After this WCO was converted into biodiesel through transesterification process. The amount of catalyst used for transesterification was determined using Equation (1)

$$\text{Catalyst amount } = \frac{\text{Catalyst concentration } \times \text{ Amount of WCO used}}{100} \tag{1}$$

In the presence of KOH catalyst and methanol, WCO has been converted into biodiesel by transesterification process [16,40]. In general, methanol is soluble in FAME or biodiesel but insoluble in triglycerides [41]. Methanol was added to WCO in the presence of KOH at the temperature range from (50 to 65 °C) with reaction times of 1–3 h which settled down overnight. Biodiesel appeared to be collected in the top layer whereas glycerine settled down in the bottom, the later was separated with a separating funnel. Transesterified biodiesel was washed with hot water continuously for removal of impurities which included catalysts and unused methanol. Biodiesel was washed using distilled water, and this process was done until the used distilled water became transparent. After this, the rotary evaporator was used to remove the remains of water content and methanol form biodiesel. After rotary evaporation, biodiesel was filtered with Whatman filter paper to remove the traces amount of KOH catalyst. Biodiesel yield was calculated using Equation (2)

$$\text{Yield} = \frac{\text{Amount of biodiesel produced}}{\text{amount of WCO used}} \times 100 \tag{2}$$

2.3. Biodiesel Property Analysis

The lower heating value (LHV) of biodiesel was determined using a bomb calorimeter. The flashpoint and fire point of biodiesel was measured using the Cleveland open cup apparatus (Koehler, New York, NY, USA). The FAME composition was determined using GCMS 5975C with triple I Detector. Helium gas was used as a carrier gas. For determination of the acid value of WCO, 0.5 N KOH was mixed with 50 mL distilled water and the mixture was used for titration. A mixture of 0.25 g phenathpelin and 25 mL ethyl alcohol was used as an indicator. A solution of 50 mL (95% ethyl alcohol and 5% distilled water) was prepared, and a 1 ml indicator was added into a solution of WCO. The AV of WCO was calculated using Equation (3) [40,42]

$$\text{Acid Value} = \frac{56.1 \times N \times V}{W} \tag{3}$$

where N is the normality of KOH, V is the volume of KOH and distilled water used for titration, and W is the weight of WCO used.

2.4. Method for Biodiesel Yield Optimisation

Five major operating parameters that affect biodiesel yield are temperature, catalyst concentration, and methanol to oil ratio, reaction speed and time. Design-Expert software 8.0.6 was used to design experimental conditions for the optimisation of biodiesel yield. The six-level, five factors CCD has been used in this study that requires 46 experiments. The ranges of operating parameters have been shown in Table 2.

Table 2. Ranges of operating parameters.

Operating Parameters	Unit	Ranges
Temperature	°C	50–65
Catalyst Concentration	*w/w*	0.25–1.75
Reaction Time	h	1–3
Reaction Speed	rpm	400–800
Methanol to oil ratio	-	5:1–12:1

The data collected from performed experiments were analysed on Design-Expert 8.0.6 (Stat-Ease, Minnesota, MN, USA) and then interpreted. Three main analytical steps required to develop optimum conditions include regression analysis, plotting of response surface and analysis of variance (ANOVA). After optimising the yield, biodiesel was produced at a pilot scale using the optimised parameters for the engine performance test. Biodiesel blends were formulated using an electrical homogeniser. This homogeniser was rotated at 1500 rpm for 20 min to mix HSD with biodiesel in different concentrations. B10, B20, B30, B40, and B50 were prepared on volume bases where the concentration of biodiesel is specified with digits in blend name.

2.5. Engine Setup

Blends of biodiesel obtained from WCO with diesel (B10, B20, B30, B40, and B50) was used to run the CI engine. The pictorial view, as well as the schematic diagram of the diesel engine test rig, is exhibited in Figure 1. The experimental setup consists of a six-cylinder, four-stroke, water-cooled, indirect injection diesel engine connected with a hydraulic dynamometer. The characteristics of the CI engine were shown in Table 3. Before collecting data, in every test, steady-state engine operation was ensured by running the engine for 20 min. Initially, the engine was operated with pure diesel to collect baseline/reference data and later on performance with biodiesel blends was tested. Engine performance (brake torque, brake power, BTE, and BSFC) at different engine speeds (1000–3500 rpm) at full load conditions were tested and reported in the following sections.

(a)

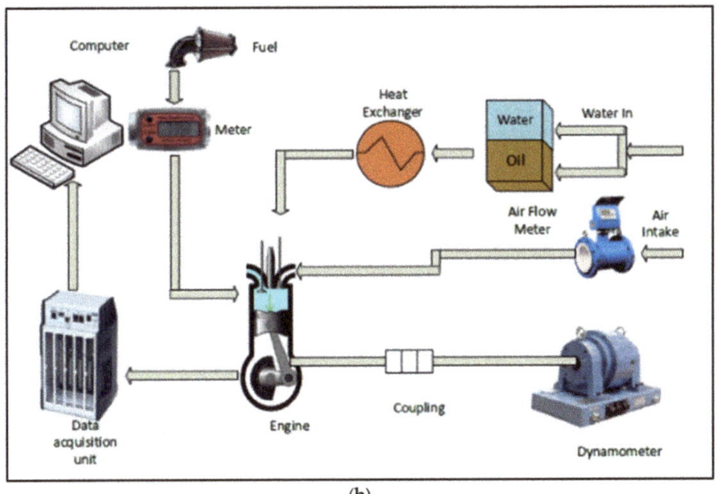

(b)

Figure 1. (**a**) Pictorial view and (**b**) schematic of the experimental testbed.

Table 3. Specification of CI engine.

Description	Specifications
No. of cylinders	6
Displacement (cm^3)	2825
Bore (mm)	85
Stroke (mm)	83
Cooling system	Water-cooled
Compression ratio	22:4
Maximum power	73.55 kW/4800 rpm
Maximum torque	178.48 Nm/3000 rpm
Dynamometer	Hydraulic
Injection system	Indirect

3. Results and Analysis

3.1. Physicochemical Properties of Biodiesel

Table 4 presents the physicochemical properties of biodiesel sourced from the WCO. These properties have been compared with standard thermos physical properties of biodiesel as per ASTM standards. GCMS has determined the components of FAME, Table 5 illustrates the percentage composition of different long carbon chain elements.

Table 4. Physicochemical characteristics of biodiesel blends and HSD.

Properties	Diesel	B100	B10	B20	B30	B40	B50
Density at 15 °C (kg/m^3)	831	892	833.5	837.5	842.5	851.5	859.5
Viscosity at 40 °C (mm^2/s)	3.9016	5.69	3.993	3.58	4.163	4.648	4.933
Acid value (mg KOH/g)	<0.247	0.6732	0.25	0.265	0.269	0.276	0.285
Flashpoint (°C)	79	140	135.78	133.34	130	124.56	120
Pour point (°C)	7	1.2	1.5	1.8	2	2.5	3
Lower Heating Value (MJ/kg)	44.2	38.753	42.209	41.987	41.632	40.098	39.456

Table 5. FAME (*w/w* %) biodiesel produced from WCO.

Common Name	Structure	WCOME
Methyl Palmitate	C16:0	6
Methyl Palmitoleate	C16:1	0.6
Methyl Eicosenoate	C20:1	0.8
Methyl Behenate	C22:0	0.9
Methyl Arachidate	C20:0	0.9
Methyl Stearate	C18:0	0.8
Methyl Oleate	C18:1	54.0
Methyl Linoleate	C18:2	25.7
Methyl Erucate	C22:1	2.2
Methyl Linolenate	C18:3	8.1

3.2. Model Fitting and ANOVA

Table 6 represents ANOVA parameters for the quadratic polynomial model. These predicted values are obtained from model fitting techniques using Design–Expert software 8.0.6. There are various data fitting models that include linear, two factorial, cubic and quadratic. The quadratic polynomial model was used for data fitting, and the equation for this model is shown below.

$$\text{Yield (\%)} = +84.33 - 0.16X_1 - 5.34X_2 - 7.59X_3 + 3.44X_4 + 0.97X_5 + 4.00X_1X_2 + 1.75X_1X_3$$
$$+ 5.50X_1X_4 + 2.13X_1X_5 + 6.38X_2X_3 - 0.38X_2X_4 - 0.13X_2X_5 - 0.38X_3X_4 + 1.62X_3X_5 + \tag{4}$$
$$0.50X_4X_5 - 2.56X_1{}^2 - 2.06X_2{}^2 - 6.81X_3{}^2 - 1.19X_4{}^2 - 2.90X_5{}^2$$

where X_1 is temperature, X_2 is catalyst concentration, X_3 is methanol to oil ratio, and X_4 is stirring speed, X_5 is reaction time. The model F-value of 2.70 means that the model is significant. A model F-value this large could only occur at 1.00% due to noise. "Prob > F" values less than 0.0500 show significant model terms. Values over 0.1000 are not significant in terms of the model. The experimental matrix developed by Design-Expert software which consists of CCD arrangements and responses has been shown in Table 7.

Table 6. ANOVA for the quadratic polynomial model.

Source	Sum of Squares	Degree of Freedom	Mean Square	F Value	Prob > F
Model	2401.89	20	120.09	2.7	0.0100 [a]
A-Temperature	0.39	1	0.39	8.78×10^{-3}	0.9261 [b]
B-Catalyst concentration	456.89	1	456.89	10.27	0.0037 [a]
C-Methanol/Oil	922.64	1	922.64	20.74	0.0001 [a]
D-stirring speed	189.06	1	189.06	4.25	0.0498 [a]
E-Time	15.02	1	15.02	0.34	0.5664 [b]
AB	64	1	64	1.44	0.2416 [b]
AC	12.25	1	12.25	0.28	0.6043 [b]
AD	121	1	121	2.72	0.1116 [b]
AE	18.06	1	18.06	0.41	0.5297 [b]
BC	162.56	1	162.56	3.66	0.0674 [a]
BD	0.56	1	0.56	0.013	0.9114 [b]
BE	0.063	1	0.063	1.41×10^{-3}	0.9704 [b]
CD	0.56	1	0.56	0.013	0.9114 [b]
CE	10.56	1	10.56	0.24	0.6303 [b]
DE	1	1	1	0.022	0.8820 [b]
A2	57.31	1	57.31	1.29	0.2671 [b]
B2	37.13	1	37.13	0.83	0.3696 [b]
C2	405.03	1	405.03	9.11	0.0058 [a]
D2	12.31	1	12.31	0.28	0.6035 [b]
E2	73.19	1	73.19	1.65	0.2113 [b]
Residual	1111.92	25	44.48		
Lack of Fit	981.58	20	49.08	1.88	0.2498 [b]
Pure Error	130.33	5	26.07		
Cor Total	3513.8	45			

[a] Significant at "prob > F" less than 0.05.; [b] Insignificant at "prob > F" more than 0.100.

Table 7. CCD arrangements and responses.

Trail No	Temperature (A)	Catalyst Concertation (B)	Methanol-Oil Ratio (C)	Stirring Speed (D)	Time (E)	Biodiesel Yield
1	65	1.75	8.5	600	2	76
2	57.5	1	5	600	3	84
3	65	1	8.5	600	3	77
4	57.5	1	5	800	2	90
5	57.5	0.25	12	600	2	63.5
6	57.5	1	8.5	800	3	85
7	57.5	1	12	800	2	76
8	65	1	5	600	2	86.5
9	65	1	12	600	2	77.5
10	57.5	1	8.5	600	2	84.5
11	57.5	1	12	600	3	65.5
12	50	1	8.5	800	2	84
13	57.5	1	8.5	600	2	91.5
14	57.5	1	5	400	2	75
15	57.5	1	5	600	1	86.5
16	57.5	1	8.5	800	1	74
17	57.5	1	8.5	600	2	75.5
18	57.5	0.25	8.5	600	1	91

Table 7. *Cont.*

Trail No	Temperature (A)	Catalyst Concertation (B)	Methanol-Oil Ratio (C)	Stirring Speed (D)	Time (E)	Biodiesel Yield
19	65	1	8.5	800	2	92.5
20	50	1	8.5	600	3	76
21	57.5	0.25	5	600	2	89.5
22	57.5	1.75	12	600	2	72.5
23	65	1	8.5	400	2	71
24	57.5	1	12	600	1	61.5
25	57.5	1.75	8.5	800	2	71
26	50	1	12	600	2	63
27	57.5	0.25	8.5	600	3	93
28	50	1	8.5	600	1	85
29	57.5	1	8.5	400	1	70
30	57.5	1	8.5	400	3	79
31	57.5	0.25	8.5	400	2	93
32	65	1	8.5	600	1	77.5
33	57.5	1.75	8.5	600	1	73
34	57.5	1	8.5	600	2	85
35	57.5	1	8.5	600	2	85
36	57.5	1.75	8.5	400	2	74
37	57.5	1.75	8.5	600	3	74.5
38	57.5	1	12	400	2	62.5
39	57.5	0.25	8.5	800	2	91.5
40	50	0.25	8.5	600	2	83.5
41	50	1	8.5	400	2	84.5
42	50	1.75	8.5	600	2	74.5
43	57.5	1.75	5	600	2	73
44	65	0.25	8.5	600	2	69
45	57.5	1	8.5	600	2	84.5
46	50	1	5	600	2	79

3.3. Effect of Operating Parameters

This section explains the effects of temperature, catalyst concentration, methanol to oil ratio, reaction stirring speed and reaction time on the biodiesel yield. Figure 2a presents experimentally obtained RSM plot to investigate the effect of catalyst concentration, methanol to oil ratio, stirring speed, reaction time, at a constant temperature range. The amount of WCO sample was 20 g, and the operating temperature ranged between 50 °C to 65 °C. The catalyst concentration was varied between 0.25 *w/w* to 1.75 *w/w*. As shown in Figure 2a, biodiesel yield increased from 50 °C to 57.50 °C and decreased afterwards. The yield decreased by increasing catalyst concentration from 0.25 *w/w* to 1.75 *w/w*. Figure 2b illustrates a relation between temperature, methanol to oil ratio and percentage yield of biodiesel at a constant catalyst concentration of 1 *w/w*, constant stirring speed of 600 rpm and constant reaction time of 2 h. The biodiesel yield was increased from 5:1 to 8:1 and decreased afterwards. The maximum yield of 93% was obtained at 8.50:1 methanol to oil ratio. Figure 2c illustrates a relationship between temperature, stirring speed and yield. Biodiesel yield was optimised at stirring speeds between 560 to 640 rpm, and it decreased at higher stirring speeds. Figure 2d illustrates a relationship between temperature, reaction time and percentage yield of biodiesel at a constant catalyst concentration of 1 *w/w*, methanol to oil ratio of 8.50:1 and a constant stirring speed of 600 rpm. A slight increase in yield was observed by increasing reaction time from 1 h to 2 h. Biodiesel yield was enhanced when the temperature increased to 57.50 °C and then decreased by increasing the temperature further. The above-stated trends are consistent with the results already reported in the literature [43,44].

Figure 3a illustrates response surface plots as a function of catalyst concentration, methanol to oil ratio and percentage yield of biodiesel at a constant temperature of 57.50 °C, the rotational speed of 600 rpm and constant reaction time of 2 h. Minimum biodiesel yield was observed at 12:1 methanol to oil ratio with catalyst concentration of 1.75 *w/w*. Maximum biodiesel yield (93%) was observed at 8.50:1 methanol to oil ratio and 0.25 *w/w* catalyst concentration. Figure 3b shows that, by changing rotational speed from 400 rpm to 800 rpm, there is a sharp increase in percentage yield from 80% to

90% at a catalyst concentration of 0.25 *w/w*. Figure 3c illustrates that both reaction time and catalyst concentration affect the percentage yield. By increasing the reaction time up to a certain limit, and by decreasing catalyst concentration, the percentage yield of biodiesel increased.

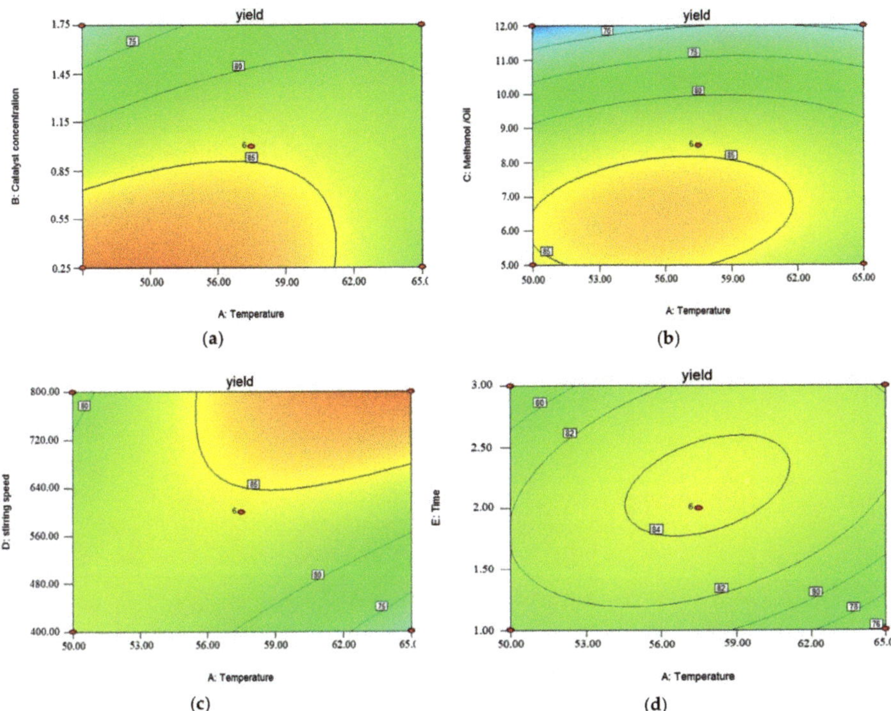

Figure 2. Experimentally obtained RSM plot to investigate the effect of (**a**) catalyst concentration, (**b**) methanol to oil ratio (**c**) stirring speed (**d**) reaction time, at a constant range of temperature.

Figure 4a illustrates response surface plots as a function of methanol to oil ratio, stirring speed and percentage yield of biodiesel at a constant temperature, catalyst concentration and reaction time that are 57.50 °C, 1 *w/w* and 2 h respectively. Percentage yield sharply increased from 70% to 90% by increasing the stirring speed from 400 rpm to 800 rpm with methanol to oil ratio 8.50:1. As shown in Figure 4b, the optimum yield of biodiesel was observed at 7:1 to 8:1 methanol to oil ratio. Reaction time also affects the yield. Ghadge et al. [45] also reported the use of the same methanol to oil ratio in his investigation. Figure 5, illustrates response surface plots as a function of stirring speed, reaction time and percentage yield of biodiesel at a constant temperature, catalyst concentration, and methanol to oil ratio that are 57.50 °C, 1 *w/w* and 8.50:1 respectively. The graph shows a steady increase in yield by increasing rotational speed and reaction time.

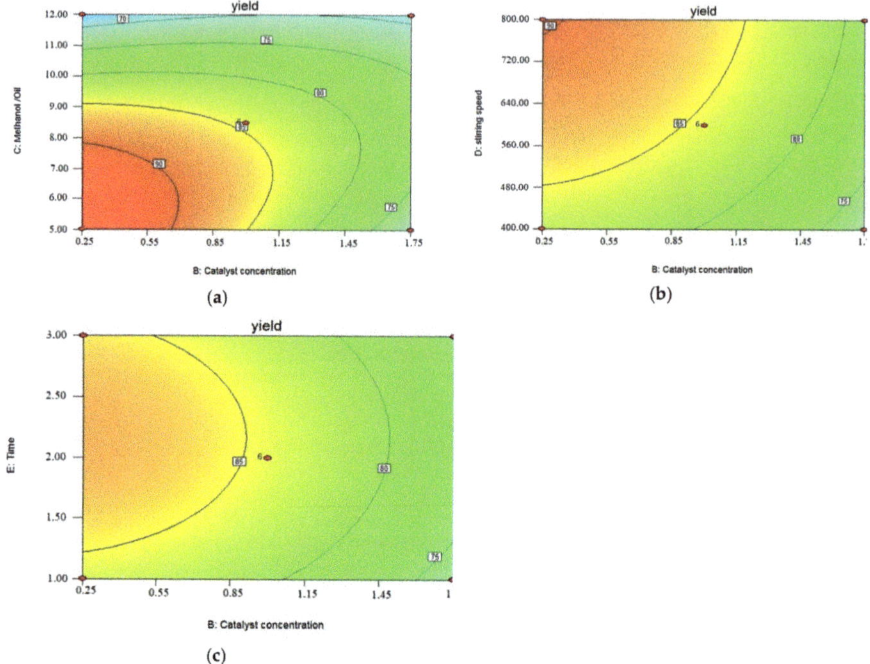

Figure 3. Experimentally obtained RSM plot to investigate the effect of (**a**) methanol to oil ratio, (**b**) stirring speed and (**c**) reaction time, at a constant range of catalyst concentration.

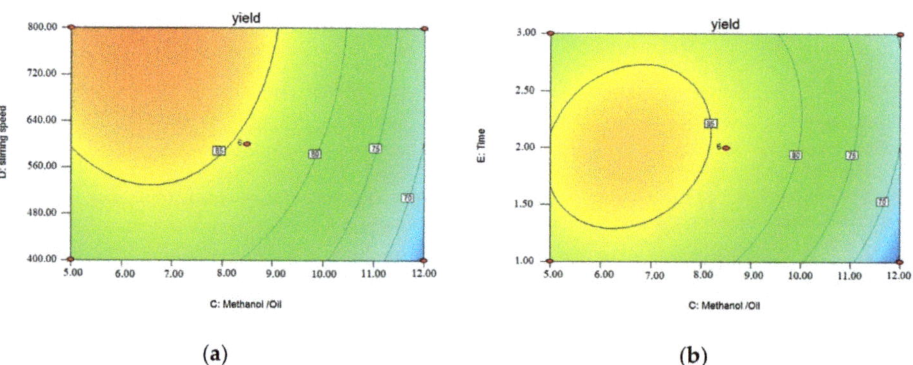

Figure 4. Experimentally obtained RSM plot to investigate the effect of (**a**) stirring speed, (**b**) reaction time, at a constant range of methanol to oil ratio.

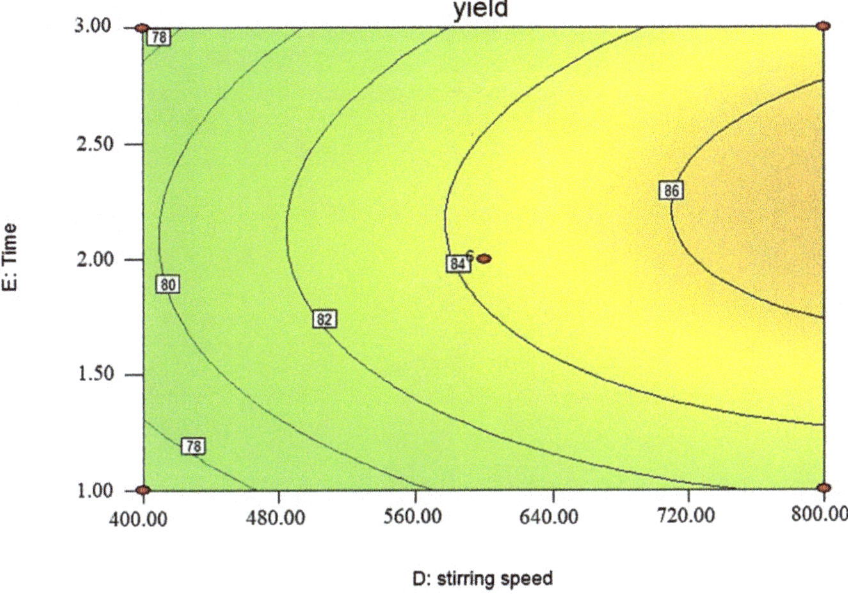

Figure 5. Experimentally obtained RSM plot to investigate the effect of reaction time at a constant range of stirring speed.

3.4. Performance Characteristics

3.4.1. Brake Torque and Brake Power

Figure 6a,b shows the variation of brake torque and brake power with engine speed at full load condition for biodiesel blends and HSD. The trends for biodiesel blends were found to be similar to the HSD. Initially, the brake torque for HSD and biodiesel blends increases with the increase in engine speed, and it reaches a maximum value and then starts to decrease afterwards. This trend can be attributed to: firstly, the volumetric efficiency of the engine which decreases with increase in engine speed and secondly, the increase in frictional losses at higher engine speeds which results in the reduction of the brake torque. The maximum values of engine torque for both HSD and biodiesel blends were found at 3000 rpm. Biodiesel blends exhibited slightly lower brake torque than that of HSD. This can be attributed to the relatively low LHV and higher kinematic viscosity of biodiesel blends. Higher kinematic viscosity results in a greater delay in the start of injection, which also leads to poorer fuel atomisation [46]. The average torque reduction for biodiesel blends compared to HSD at 3000 rpm was found to be 1.45% for B10, 2% for B20, 2.2% for B30, 3.09% for B40, and 3.5% for B50. Brake power is derived from brake torque by multiplying with the angular speed. From Figure 6b, the highest power was observed at 3500 rpm. Again, the highest power was shown by HSD followed by B10, B20, B30, B40 and B50, respectively. Rizwanul Fattah et al. [46] studied the engine performance, and emission characteristics of Malaysian Alexandrian Laurel oil-based biodiesel (ALB) blend with diesel. They also reported a slight reduction in maximum brake powers for ALB10 and ALB20 compared to that of diesel fuel. The power outputs were 48.5 kW, 48.3 kW, and 48.2 kW, for diesel, ALB10, and ALB20, respectively. They attributed this reduction to the lower LHV and higher viscosity of blends, both of which result in poor fuel atomisation compared to that of diesel fuel.

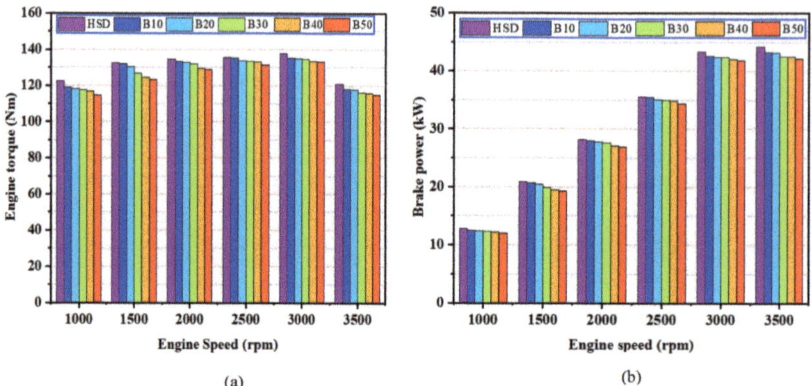

Figure 6. Effect of different fuels on (**a**) BT and (**b**) BP of the engine at different speeds.

3.4.2. Brake Specific Fuel Consumption and Brake Thermal Efficiency

Figure 7a presents the variation in BSFC for biodiesel blends and HSD as a function of engine speed. Biodiesel blends have lower LHV and higher densities compared to that of HSD, resulting in higher fuel consumption for biodiesel blends. It is evident for the figure that, BSFC first decreased to the lowest at 2000 rpm, and then increased with increase in speed. The average increase in BSFC for biodiesel blends as compared to HSD at 3500 rpm is found as 1.61% for B10, 5.73% for B20, 8.8% for B30, 12.76% for B40, and 18% for B50. The increase in BSFC for biodiesel blends was due to volumetric effect of the constant fuel injection rate together with the higher kinematic viscosity and lower LHV of biodiesel and its blends, and this became more pronounced in higher biodiesel blends such as B50. Palash et al. [47] also reported a similar finding for *Aphanamixis polystachya* biodiesel (APME) blends with the average BSFC values for diesel, APME5, and APME10 of 352.96 g/kWh, 356.05 g/kWh, and 359.29 g/kWh, respectively. Figure 7b represents variation in BTE as a function of engine speed at full load conditions. Thermal efficiency varies inversely with the product of BSFC and LHV of the fuel [46]. The highest BTE for all fuels were observed at 3000 rpm at full load condition. The BTE values for HSD, B10, B20, B30, B40 and B50 are 34.97%, 32.78%, 33.36%, 33.76%, 34.15%, and 32.11%, respectively.

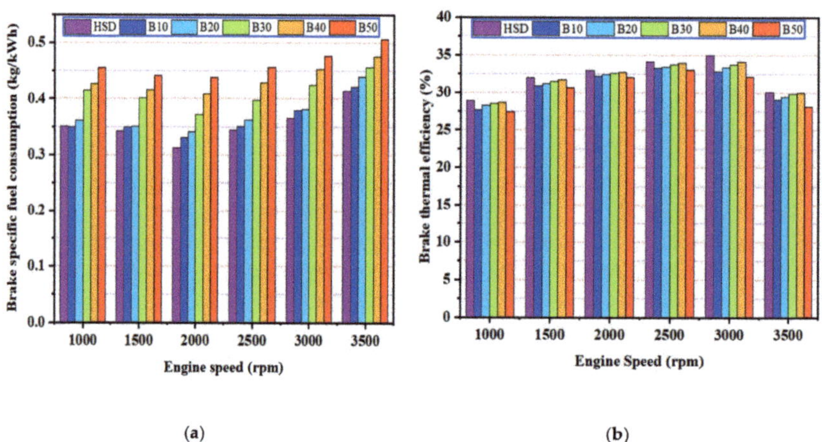

Figure 7. Effect of different fuels on (**a**) BSFC and (**b**) BTE of the engine at different speeds.

4. Conclusions

Biodiesel yield was optimised at optimum operating conditions. FFAs of WCO were reduced by acid treatment. Among mineral acids, H_2SO_4 was found to be more effective. It reduces the FFAs value of 91.36%. Transesterification of WCO with methanol was found to be very effective. A biodiesel yield of 93% was obtained with a 57.50 °C reaction temperature, 0.25% catalyst concentration, 8.50:1 methanol to oil ratio, 600 rpm stirring speed, and 3 h reaction time. Various physiochemical properties justify the quality of biodiesel. Oleic acid (C18:1), linoleic acid (C18:2), α-linoleic acid (C18:3), and palmitic acid (C16:0) were the major constituents of biodiesel. Performance characteristics of a diesel engine using diesel fuel and biodiesel blends B10, B20, B30, B40, and B50 were tested and compared. Engine performance in terms of brake torque, brake power, BSFC, and BTE was determined. The average torque reduction for biodiesel blends compared to HSD at 3000 rpm were found to be 1.45%, 2%, 2.2%, 3.09%, and 3.5% for B10, B20, B30, B40, and B50, respectively. The average increase in BSFC for biodiesel blends compared to HSD at 3500 rpm were found to be 1.61%, 5.73%, 8.8%, 12.76%, and 18% for B10, B20, B30, B40, and B50, respectively. Pakistan is spending a huge amount of money in importing HSD to fulfil their energy requirements. The government is looking for alternative fuels that have the potential to reduce the consumption of HSD. At present, the most suitable alternative fuel is biodiesel produced from WCO. The conversion of WCO into biodiesel not only reduces the import of HSD, but also reduces environmental concerns.

Author Contributions: Conceptualisation, L.R., Z.A., and S.I.; methodology, M.E.M.S.; software, M.E.M.S.; validation, M.M.A., H.M.K., and T.A.; formal analysis, M.M.A.; investigation N.S.; resources, M.M.A. and M.E.M.S.; data curation, M.E.M.S.; writing—original draft preparation, M.F.; writing—review and editing, S.M.A.R., I.M.R.F., and M.M.A.; visualisation, M.A.; supervision, L.R.; project administration, N.S.; funding acquisition, M.M.A. All authors have read and agreed to the published version of the manuscript.

Funding: This research received no external funding.

Acknowledgments: The authors would like to thank to the Pakistan Council of Scientific and Industrial Research (PCSIR) for their services regarding biodiesel characterisation. Authors would also like to acknowledge 'research development fund' of School of Information, Systems and Modelling, Faculty of Engineering and Information Technology, University of Technology Sydney, Australia.

Conflicts of Interest: The authors declare no conflict of interest.

Nomenclature

ANOVA	Analysis of Variance
ANFIS	Adaptive neuro-fuzzy inference system
AV	Acid value
BSFC	Brake specific fuel consumption
BP	Brake power
BTE	Brake thermal efficiency
BT	Brake torque
CCD	Central Composite Design
CI	Compression ignition
DI	Direct injection
FAME	Fatty acid methyl ester
FFA	Free fatty acid
GCMS	Gas chromatography-mass spectroscopy
HSD	High-speed diesel
HPLC	High-Performance Liquid Chromatography
LHV	Lower heating value
NMR	Nuclear magnetic resonance
RSM	Response surface methodology
WCO	Waste cooking oil

References

1. Soudagar, M.E.M.; Banapurmath, N.R.; Afzal, A.; Hossain, N.; Abbas, M.M.; Haniffa, M.A.C.M.; Naik, B.; Ahmed, W.; Nizamuddin, S.; Mubarak, N.M. Study of diesel engine characteristics by adding nanosized zinc oxide and diethyl ether additives in Mahua biodiesel–diesel fuel blend. *Sci. Rep.* **2020**, *10*, 15326. [CrossRef] [PubMed]

2. Hussain, F.; Soudagar, M.E.M.; Afzal, A.; Mujtaba, M.; Fattah, I.M.R.; Naik, B.; Mulla, M.H.; Badruddin, I.A.; Khan, T.M.Y.; Raju, V.D.; et al. Enhancement in Combustion, Performance, and Emission Characteristics of a Diesel Engine Fueled with Ce-ZnO Nanoparticle Additive Added to Soybean Biodiesel Blends. *Energies* **2020**, *13*, 4578. [CrossRef]

3. Mujtaba, M.A.; Masjuki, H.H.; Kalam, M.A.; Noor, F.; Farooq, M.; Ong, H.C.; Gul, M.; Soudagar, M.E.M.; Bashir, S.; Fattah, I.M.R.; et al. Effect of Additivized Biodiesel Blends on Diesel Engine Performance, Emission, Tribological Characteristics, and Lubricant Tribology. *Energies* **2020**, *13*, 3375. [CrossRef]

4. Baudry, G.; Macharis, C.; Vallée, T. Can microalgae biodiesel contribute to achieve the sustainability objectives in the transport sector in France by 2030? A comparison between first, second and third generation biofuels though a range-based Multi-Actor Multi-Criteria Analysis. *Energy* **2018**, *155*, 1032–1046. [CrossRef]

5. Usman, M.; Farooq, M.; Naqvi, M.; Saleem, M.W.; Hussain, J.; Naqvi, S.R.; Jahangir, S.; Jazim Usama, H.M.; Idrees, S.; Anukam, A. Use of gasoline, LPG and LPG-HHO blend in SI engine: A comparative performance for emission control and sustainable environment. *Processes* **2020**, *8*, 74. [CrossRef]

6. Masjuki, H.H.; Ruhul, A.M.; Mustafi, N.N.; Kalam, M.A.; Arbab, M.I.; Rizwanul Fattah, I.M. Study of production optimization and effect of hydroxyl gas on a CI engine performance and emission fueled with biodiesel blends. *Int. J. Hydrog. Energy* **2016**, *41*, 14519–14528. [CrossRef]

7. Shahir, S.A.; Masjuki, H.H.; Kalam, M.A.; Imran, A.; Rizwanul Fattah, I.M.; Sanjid, A. Feasibility of diesel–biodiesel–ethanol/bioethanol blend as existing CI engine fuel: An assessment of properties, material compatibility, safety and combustion. *Renew. Sustain. Energy Rev.* **2014**, *32*, 379–395. [CrossRef]

8. Shuba, E.S.; Kifle, D. Microalgae to biofuels: 'Promising' alternative and renewable energy, review. *Renew. Sustain. Energy Rev.* **2018**, *81*, 743–755. [CrossRef]

9. Mahlia, T.M.I.; Syazmi, Z.; Mofijur, M.; Abas, A.E.P.; Bilad, M.R.; Ong, H.C.; Silitonga, A.S. Patent landscape review on biodiesel production: Technology updates. *Renew. Sustain. Energy Rev.* **2020**, *118*. [CrossRef]

10. Serin, H.; Yıldızhan, Ş. Hydrogen addition to tea seed oil biodiesel: Performance and emission characteristics. *Int. J. Hydrog. Energy* **2018**, *43*, 18020–18027. [CrossRef]

11. Imran, S.; Korakianitis, T.; Shaukat, R.; Farooq, M.; Condoor, S.; Jayaram, S. Experimentally tested performance and emissions advantages of using natural-gas and hydrogen fuel mixture with diesel and rapeseed methyl ester as pilot fuels. *Appl. Energy* **2018**, *229*, 1260–1268. [CrossRef]

12. Imran, S.; Emberson, D.R.; Hussain, A.; Ali, H.; Ihracska, B.; Korakianitis, T. Performance and specific emissions contours throughout the operating range of hydrogen-fueled compression ignition engine with diesel and RME pilot fuels. *Alex. Eng. J.* **2015**, *54*, 303–314. [CrossRef]

13. Alagu, K.; Venu, H.; Jayaraman, J.; Raju, V.D.; Subramani, L.; Appavu, P.; Dhanasekar, S. Novel water hyacinth biodiesel as a potential alternative fuel for existing unmodified diesel engine: Performance, combustion and emission characteristics. *Energy* **2019**, *179*, 295–305. [CrossRef]

14. Vigneswaran, R.; Annamalai, K.; Dhinesh, B.; Krishnamoorthy, R. Experimental investigation of unmodified diesel engine performance, combustion and emission with multipurpose additive along with water-in-diesel emulsion fuel. *Energy Convers. Manag.* **2018**, *172*, 370–380. [CrossRef]

15. Yunus Khan, T.M.; Badruddin, I.A.; Ankalgi, R.F.; Badarudin, A.; Hungund, B.S.; Ankalgi, F.R. Biodiesel Production by Direct Transesterification Process via Sequential Use of Acid–Base Catalysis. *Arab. J. Sci. Eng.* **2018**, *43*, 5929–5936. [CrossRef]

16. Fattah, I.M.R.; Ong, H.C.; Mahlia, T.M.I.; Mofijur, M.; Silitonga, A.S.; Rahman, S.M.A.; Ahmad, A. State of the art of catalysts for biodiesel production. *Front. Energy Res.* **2020**. [CrossRef]

17. Vicente, G.; Martínez, M.; Aracil, J. Integrated biodiesel production: A comparison of different homogeneous catalysts systems. *Bioresour. Technol.* **2004**, *92*, 297–305. [CrossRef]

18. Mekala, M.; Goli, V.R. Kinetics of esterification of methanol and acetic acid with mineral homogeneous acid catalyst. *Chin. J. Chem. Eng.* **2015**, *23*, 100–105. [CrossRef]

19. Demirbas, A. Biodiesel production from vegetable oils via catalytic and non-catalytic supercritical methanol transesterification methods. *Prog. Energy Combust. Sci.* **2005**, *31*, 466–487. [CrossRef]
20. Likozar, B.; Levec, J. Effect of process conditions on equilibrium, reaction kinetics and mass transfer for triglyceride transesterification to biodiesel: Experimental and modeling based on fatty acid composition. *Fuel Process. Technol.* **2014**, *122*, 30–41. [CrossRef]
21. Ahmad, T.; Danish, M.; Kale, P.; Geremew, B.; Adeloju, S.B.; Nizami, M.; Ayoub, M. Optimization of process variables for biodiesel production by transesterification of flaxseed oil and produced biodiesel characterizations. *Renew. Energy* **2019**, *139*, 1272–1280. [CrossRef]
22. Klofutar, B.; Golob, J.; Likozar, B.; Klofutar, C.; Žagar, E.; Poljanšek, I. The transesterification of rapeseed and waste sunflower oils: Mass-transfer and kinetics in a laboratory batch reactor and in an industrial-scale reactor/separator setup. *Bioresour. Technol.* **2010**, *101*, 3333–3344. [CrossRef] [PubMed]
23. Fattah, I.M.R.; Yip, H.L.; Jiang, Z.; Yuen, A.C.Y.; Yang, W.; Medwell, P.R.; Kook, S.; Yeoh, G.H.; Chan, Q.N. Effects of flame-plane wall impingement on diesel combustion and soot processes. *Fuel* **2019**, *255*, 115726. [CrossRef]
24. Yip, H.L.; Rizwanul Fattah, I.M.; Yuen, A.C.Y.; Yang, W.; Medwell, P.R.; Kook, S.; Yeoh, G.H.; Chan, Q.N. Flame–Wall Interaction Effects on Diesel Post-injection Combustion and Soot Formation Processes. *Energy Fuels* **2019**, *33*, 7759–7769. [CrossRef]
25. Carraretto, C.; Macor, A.; Mirandola, A.; Stoppato, A.; Tonon, S. Biodiesel as alternative fuel: Experimental analysis and energetic evaluations. *Energy* **2004**, *29*, 2195–2211. [CrossRef]
26. Rashedul, H.K.; Masjuki, H.H.; Kalam, M.A.; Ashraful, A.M.; Ashrafur Rahman, S.M.; Shahir, S.A. The effect of additives on properties, performance and emission of biodiesel fuelled compression ignition engine. *Energy Convers. Manag.* **2014**, *88*, 348–364. [CrossRef]
27. Ashraful, A.M.; Masjuki, H.H.; Kalam, M.A.; Rizwanul Fattah, I.M.; Imtenan, S.; Shahir, S.A.; Mobarak, H.M. Production and comparison of fuel properties, engine performance, and emission characteristics of biodiesel from various non-edible vegetable oils: A review. *Energy Convers. Manag.* **2014**, *80*, 202–228. [CrossRef]
28. Nirmala, N.; Dawn, S.S.; Harindra, C. Analysis of performance and emission characteristics of Waste cooking oil and *Chlorella variabilis* MK039712.1 biodiesel blends in a single cylinder, four strokes diesel engine. *Renew. Energy* **2020**, *147*, 284–292. [CrossRef]
29. Akcay, M.; Yilmaz, I.T.; Feyzioglu, A. Effect of hydrogen addition on performance and emission characteristics of a common-rail CI engine fueled with diesel/waste cooking oil biodiesel blends. *Energy* **2020**, *212*, 118538. [CrossRef]
30. Can, Ö. Combustion characteristics, performance and exhaust emissions of a diesel engine fueled with a waste cooking oil biodiesel mixture. *Energy Convers. Manag.* **2014**, *87*, 676–686. [CrossRef]
31. Nayak, M.G.; Vyas, A.P. Optimization of microwave-assisted biodiesel production from Papaya oil using response surface methodology. *Renew. Energy* **2019**, *138*, 18–28. [CrossRef]
32. Mostafaei, M.; Javadikia, H.; Naderloo, L. Modeling the effects of ultrasound power and reactor dimension on the biodiesel production yield: Comparison of prediction abilities between response surface methodology (RSM) and adaptive neuro-fuzzy inference system (ANFIS). *Energy* **2016**, *115*, 626–636. [CrossRef]
33. Mubarak, M.; Shaija, A.; Suchithra, T.V. Optimization of lipid extraction from Salvinia molesta for biodiesel production using RSM and its FAME analysis. *Environ. Sci. Pollut. Res.* **2016**, *23*, 14047–14055. [CrossRef] [PubMed]
34. Mehta, K.; Divya, N.; Jha, M.K. Application of RSM for Optimizing the Biodiesel Production Catalyzed by Calcium Methoxide. In *Sustainable Engineering*; Springer: Singapore, 2019; pp. 75–83.
35. Latchubugata, C.S.; Kondapaneni, R.V.; Patluri, K.K.; Virendra, U.; Vedantam, S. Kinetics and optimization studies using Response Surface Methodology in biodiesel production using heterogeneous catalyst. *Chem. Eng. Res. Des.* **2018**, *135*, 129–139. [CrossRef]
36. Jamshaid, M.; Masjuki, H.H.; Kalam, M.A.; Zulkifli, N.W.M.; Arslan, A.; Alwi, A.; Khuong, L.S.; Alabdulkarem, A.; Syahir, A.Z. Production optimization and tribological characteristics of cottonseed oil methyl ester. *J. Clean. Prod.* **2019**, *209*, 62–73. [CrossRef]
37. Anwar, M.; Rasul, M.G.; Ashwath, N.; Rahman, M.M. Optimisation of Second-Generation Biodiesel Production from Australian Native Stone Fruit Oil Using Response Surface Method. *Energies* **2018**, *11*, 2566. [CrossRef]

38. Veljković, V.B.; Veličković, A.V.; Avramović, J.M.; Stamenković, O.S. Modeling of biodiesel production: Performance comparison of Box–Behnken, face central composite and full factorial design. *Chin. J. Chem. Eng.* **2019**, *27*, 1690–1698. [CrossRef]

39. Chai, M.; Tu, Q.; Lu, M.; Yang, Y.J. Esterification pretreatment of free fatty acid in biodiesel production, from laboratory to industry. *Fuel Process. Technol.* **2014**, *125*, 106–113. [CrossRef]

40. Fattah, I.M.R.; Masjuki, H.H.; Kalam, M.A.; Wakil, M.A.; Ashraful, A.M.; Shahir, S.A. Experimental investigation of performance and regulated emissions of a diesel engine with *Calophyllum inophyllum* biodiesel blends accompanied by oxidation inhibitors. *Energy Convers. Manag.* **2014**, *83*, 232–240. [CrossRef]

41. Santos, T.; Gomes, J.F.; Puna, J. Liquid-liquid equilibrium for ternary system containing biodiesel, methanol and water. *J. Environ. Chem. Eng.* **2018**, *6*, 984–990. [CrossRef]

42. Fattah, I.M.R.; Masjuki, H.H.; Kalam, M.A.; Mofijur, M.; Abedin, M.J. Effect of antioxidant on the performance and emission characteristics of a diesel engine fueled with palm biodiesel blends. *Energy Convers. Manag.* **2014**, *79*, 265–272. [CrossRef]

43. Balasubramaniam, B.; Sudalaiyadum Perumal, A.; Jayaraman, J.; Mani, J.; Ramanujam, P. Comparative analysis for the production of fatty acid alkyl esterase using whole cell biocatalyst and purified enzyme from *Rhizopus oryzae* on waste cooking oil (sunflower oil). *Waste Manag.* **2012**, *32*, 1539–1547. [CrossRef] [PubMed]

44. Verma, P.; Sharma, M.P. Comparative analysis of effect of methanol and ethanol on Karanja biodiesel production and its optimisation. *Fuel* **2016**, *180*, 164–174. [CrossRef]

45. Ghadge, S.V.; Raheman, H. Biodiesel production from mahua (*Madhuca indica*) oil having high free fatty acids. *Biomass Bioenergy* **2005**, *28*, 601–605. [CrossRef]

46. Fattah, I.M.R.; Kalam, M.A.; Masjuki, H.H.; Wakil, M.A. Biodiesel production, characterization, engine performance, and emission characteristics of Malaysian *Alexandrian laurel* oil. *RSC Adv.* **2014**, *4*, 17787–17796. [CrossRef]

47. Palash, S.M.; Masjuki, H.H.; Kalam, M.A.; Atabani, A.E.; Rizwanul Fattah, I.M.; Sanjid, A. Biodiesel production, characterization, diesel engine performance, and emission characteristics of methyl esters from *Aphanamixis polystachya* oil of Bangladesh. *Energy Convers. Manag.* **2015**, *91*, 149–157. [CrossRef]

Publisher's Note: MDPI stays neutral with regard to jurisdictional claims in published maps and institutional affiliations.

Article

Coal Combustion Products Management toward a Circular Economy—A Case Study of the Coal Power Plant Sector in Poland

Agnieszka Bielecka [1] and Joanna Kulczycka [2,3,*] 

1   PGE Energia Ciepła S.A., 31-587 Cracow, Poland; a.bielecka@interia.pl
2   Faculty of Management, AGH University of Science and Technology, 30-076 Cracow, Poland
3   Mineral and Energy Economy Research Institute Polish Academy of Sciences, 31-261 Cracow, Poland
*   Correspondence: kulczycka@meeri.pl; Tel.: +48-605-333-363

Received: 15 June 2020; Accepted: 9 July 2020; Published: 13 July 2020

Abstract: Coal combustion products can be considered as commercial products or waste depending on the quality of the coal, the combustion process, and the country's legislation. The circular economy can create incentives for the implementation of new business models in large power plants in cooperation with coal mines and users of coal combustion products. This is particularly important in Poland, where coal still remains the main source of energy, employing over 80,000 workers. The objective of this study was to assess the readiness for change toward a circular economy and to identify challenges, barriers, and plans at seven large power plants. To do this, a final questionnaire was developed after checking environmental reporting, a CATI survey, and brainstorming between circular economy leaders from science, industry, and non-governmental organizations. The results indicate that even if the great economic and environmental potential of coal combustion products management are understood, all requirements connected with CO_2 and air pollution have higher priorities. Policy shifts away from coal do not promote cooperation, but the higher acceptance of products from waste and more transparent data shows a large potential for changes toward a circular economy.

Keywords: circular economy; coal power plant; coal combustion products; industrial waste

---

1. Introduction

The production volumes of coal combustion products (CCPs) are directly correlated with coal combustion in thermal power stations. Their quality, however, depends on the mineral composition of the coal, combustion efficiency, type and fineness of the coal, time of storage of the minerals in the furnace/boiler, etc. [1]. They constitute major inputs in the manufacture of construction materials and geotechnical engineering applications. Some of them i.e., pulverized fly ash, are a valuable waste residue used in concrete production [2]. However, they can also pose serious threats to air, water, and soil; over the past two decades, research has been developing methods to combine both the quality of power generation, economic viability, environmental safety, quality, and the effective use of waste products [3].

The beneficial use of CCPs varies in different countries; for example, Australia reached 5.936 million tonnes or 47% of CCPs in 2018 [4]. To facilitate precision and consistency, the members of the World Wide Coal Combustion Products Network have worked together to harmonize terminology and promote CCPs as a coherent nomenclature. This terminology is more positive and is in line with the concept of industrial ecology and circular economy (CE) principles, which is an approach that aims to reuse one industry's byproducts as another industry's raw material [5].

It is estimated that approximately 750 million tonnes of coal ash are generated every year in the coal industry throughout the world. A significant volume of coal fly ash is produced by major industrial countries, where it is one of the largest single sources of industrial solid waste [6].

Utilization rates vary considerably between countries, with the highest effective utilization rate of 99.3% being reported in Japan, while Africa/Middle East (still) has the lowest rate of 10.6%. In Poland, approximately 20 million tonnes of CCPs were generated (Table 1), of which approximately 65% was utilized in 2016 and 35% was landfilled. CCPs can substitute natural aggregate, as they are mainly used for in cement industry, in construction of new permanent disposal facilities, mining applications, in structural fills, waste stabilization, gypsum panel production, and agricultural applications. These applications can reduce greenhouse gas emissions, thus increasing energy efficiency and minimize use of primary resources, which is in line with CE strategy.

Based on the example of CCPs, Park and Chertow [7] have proposed the resource potential indicator (RPI) to promote the importance of managing wastes as resources in a closed loop. This measures the intrinsic value that a material has for reuse taking into account the technological recycling availability, economic feasibility and environmental requirements for the US market.

Table 1. Amount of coal combustion products (CCPs) generated in Poland in 2016 in enterprises and energy groups [8].

Energy Group/Enterprise	Amount of CCPs [Tonne/Year]
CEZ Produkty Energetyczne	260,000
ECO S.A.	70,350
EDF Polska S.A.	1,465,800
ENEA	1,351,504
ENERGA	357,709
Fortum Power & Heat	37,000
GDF Suez	1,117,200
Grupa AZOTY	322,500
PAK	2,454,342
PGE GKiK	9,792,071
PGNiG Termika	700,002
Tauron Ciepło	238,399
Tauron Wytwarzanie	1,656,555
Veolia	473,686
TOTAL:	20,297,118

The strategies of enterprises from the coal power sector in Poland contain provisions that prove that these enterprises carry out activities fulfilling CE objectives, but they do not directly refer to the CE aspects, as most data are available in corporate social responsibility (CSR) reports.

However, the current way of presenting this information does not allow for the comparison of data. In the context of CCPs, there are also no provisions proving the tendency (required by the CE) to carry out activities aimed at eliminating the generation of waste, hence only part of it is managed. Moreover, there is neither sufficient transparency in the statistics nor clear rules concerning waste, byproducts, secondary material, etc. [9]. Therefore, the management of CCPs is not the most important goal for either Polish companies or Polish energy policy. They first have to meet air quality standards and try to reduce GHG emissions. As Hopkins and Purnell (2019) indicate, the politics of GHG allocation may hide a lack of progress towards a CE [10].

In Poland, according to government documents, i.e., Polish Energy Policy to 2040 (2018) [11] and the Polish National Energy and Climate Plan for the years 2021–2030 published in 2019 [12], it was stated that the share of coal in electricity generation will be systematically reduced. In 2030, it will reach the level of 56–60%, and in 2040, the downward trend will be maintained. However, it was underlined that there is great potential for the better utilization of CCPs as building materials and for a reduction in the use of primary raw materials, which is in line with a CE. It is also estimated that the renewable energy sources share in electricity production will increase to approximately 32% in 2030. These changes involve the most significant challenges for the current Polish energy sector relating to more stringent standards under the Industrial Emissions Directive (IED) [13] and the best available techniques (BAT) for the reduction of GHG and SO_2, NOx, and dust emissions as well as the monitoring of hydrogen chloride, hydrogen fluoride, mercury, and ammonia. Enterprises have been allowed until 16 August 2021 to adapt to the new regulations, but the adaptation of installations to meet the stricter requirements involves the need for investment. Such solutions can have positive or negative impact on the quality of CCPs. Therefore, determining the impact of techniques focusing on reducing dust and gas emissions on the quality of CCPs, as well as possibilities of their further physicochemical processing according to end-user needs, is of extreme importance [14]. One of the good examples of this is the production of synthetic gypsum from flue gas desulphurization installations. Good quality of physical and chemical parameters of gypsum produced in the Bełchatów power plant was confirmed by laboratory tests. Therefore, about 60% of gypsum in Poland is used in different sectors, mainly for the construction market in the creation of various types of prefabricated products based on synthetic gypsum [15].

Several reviews focusing on technological issues relating to CCP utilization and application have already been published [16,17]. These analyze the economic and environmental constraints and benefits [18,19]. There is, however, a lack of publications focusing on surveys of energy companies on incentives and barriers that could have an impact on decisions on managing waste as a resource according to the CE concept. Moreover, there is lack of standards on how to measure micro-level circularity [20]. Therefore, the aim of this paper is to verify whether coal power plants were ready to implement CE models for CPP management. The research carried out in 2018 aimed to

1. identify the level of knowledge of CE principles in coal power sector companies,
2. determine areas that stimulate the implementation of a CE and allow one to specify activities for its further development,
3. identify the possibility of managing more CCPs in relation to the plans for implementing the CE roadmap in Poland adopted in September 2019,
4. introduce CE (NSS 7–Circular economy–water, fossil raw materials, waste) as a national smart specialization in Poland in January 2019 [21].

An analysis of good practice and research conducted (patents) relating to CCP management in Poland shows the existence of a large potential on the supply side, but long-term actions stimulating and legal solutions facilitating cooperation within the value chain are still lacking.

2. Materials and Methods

In order to investigate the role of CE in LPP (large power plant), we follow Gonenc and Scholten [22], who suggested that future studies should take into account corporate intentions and actions taken which are in line with such intentions. We focus on CE, but in 2018, there were no LPPs that had a CE strategy or any indicators focusing on CE. Therefore, we first chose the most important LPPs and then searched the official corporate web sites of LPPs concerning environmental aims and data presented in CSR reports. This allowed us to identify the activities focusing on closing material loops and waste minimization that are in line with environmental strategies. Second, we selected key activities that can reflect on LPP actions directed toward implementing a CE and identified the managers responsible for its application. We contacted them by email or phone to ask about the role of the CE and the scope and responsibilities in the companies. Third, we chose 35 managers from different departments at seven LPPs who were potential leaders of the CE. Fourth, based on a literature review of surveys of the CE and after brainstorming with CE leaders from science and industry, a survey questionnaire was constructed and sent to industry representatives (Figure 1).

The research was carried out on a group of the seven largest capital groups in the Polish energy sector (with coal-fired plants), for each of which the share in the volume of electricity introduced into the grid amounted to more than 2% (73% in total)—PGE Polska Grupa Energetyczna S.A. (36%), Tauron Polska Energia S.A. (10%), Enea S.A. (9%), PAK S.A. (6%), EDF (8%), CEZ (2%), and PGNiG (2%) [5]. At the same time, these are the largest manufacturers of CPPs. The research was conducted with the participation of 27 managers who, due to the scope of their duties, had the knowledge necessary to carry out the research.

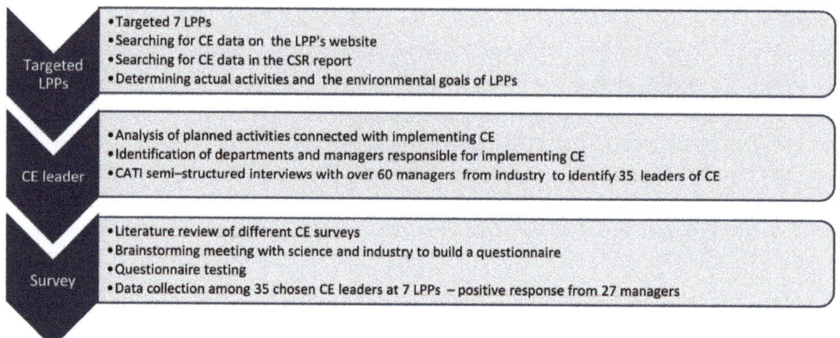

Figure 1. The circular economy (CE) implementation process.

The target group was selected from the above-mentioned companies after initially verifying the competences of the employees in departments such as Environmental Protection, Development and Quality Assurance, Development (Investments and Projects), R&D, Raw Materials and Waste, and Technological Processes. Telephone interviews (CATI) were then carried out over a period of five weeks (May–June 2018) using a questionnaire based on a five-step Likert scale (where 5 was equivalent to "strongly agree" and 1 was assigned to "strongly disagree"). The results of the research were aggregated before being presented, and the reliability of the survey was tested using the Cronbach alpha coefficient.

Aspects discussed in the survey were concerned with six topics as follows:

1. Perceived benefits relating to the potential implementation of CE;
2. Challenges related to the implementation of CE;
3. Hindrances to the implementation of CE;
4. What changes in company management are required in order to implement CE;
5. Reasons for implementing CE;
6. Prospects for implementing CE.

The detailed survey with questions is presented in Table 2. The questions (i.e., p 1_1) were chosen based on literature, reports, and experts experience and advice.

Table 2. Summary presentation of the aspects assessed in the questionnaire.

Aspect 1	Environmental Benefits (Effects) Related to CE Implementation
p 1_1	increase in efficiency of resource use
p 1_2	extended product life cycle [1]
p 1_3	introducing more sustainable production
p 1_4	reduction of environmental footprint
p 1_5	establishing less energy–intensive relations with suppliers
p 1_6	establishing relations generating less environmental pressure with consumers
p 1_7	generating products instead of waste
p 1_8	improvement in environment quality
p 1_9	strengthening environmental management standards
p 1_10	increase in ecological activity
p 1_11	closing CCPs landfills
p 1_12	reduction of the environmental cost of production

Table 2. *Cont.*

Aspect 2	Risks/Challenges Related to Implementing CE
p 2_1	increased production costs
p 2_2	need for costly and risky investments
p 2_3	need to find a way to change the process's business model
p 2_4	risk of reduced competitiveness of a company
p 2_5	need to leave the "plant gates" [2]
Aspect 3	**Barriers/Hindrances to Implementing CE**
p 3_1	failing to implement a multi–product power engineering strategy
p 3_2	failing to promote CCPs as a fully–fledged waste product
p 3_3	lack of purchasers for raw materials recovered from waste
p 3_4	lack of indicators and guidelines for implementing and measuring CE
p 3_5	current CE–consistent activities are not appreciated and evaluated
Aspect 4	**Requirements of CE with Regard to Company Management**
p 4_1	environmental responsibility going beyond the plant gates
p 4_2	cooperation within clusters
p 4_3	construction of revenue generation logic based on identification of the value chain
p 4_4	internalisation of external environmental costs
p 4_5	evaluation of environmental policy as regards suppliers and consumers
p 4_6	creation of buyers–eco–consumers' market
p 4_7	implementation of a multi–product and waste–free strategy for power engineering
p 4_8	recognition of factors representing environmental costs in the expenditure structure
p 4_9	influencing the awareness and actions of other bodies
p 4_10	identification of the value chain to specify quantifiable environmental benefits
p 4_11	integration of environmental policy with development strategy
Aspect 5	**Reasons for Implementing CE**
p 5_1	awareness of the importance of environmental protection
p 5_2	possibility of waste management cost reductions
p 5_3	social benefits
p 5_4	possible reduction of the product's environmental footprint
p 5_5	economic benefits
Aspect 6	**Assessment of the Feasibility of Implementing CE**
p 6_1	within the next 10 years
p 6_2	within the next 20 years

[1] through activities that will be carried out aimed at eliminating the storage of CCPs as a result of solutions enabling their further use in the economy, so that the resources used will not lose their value and their useful life will be extended. [2] according to the "cradle to cradle" approach.

Verification of the Research Tool

The test tool used in the research has been checked for reliability. For this purpose, an analysis of the results obtained was carried out using the Cronbach α coefficient (Table 3). This is a coefficient for the total scale, whose items can be measured on any order scale. The Cronbach α coefficient indicates what part of the variance of the total scale is the variance of the true value of that scale. A zero value means that individual scale items do not measure the true result but only generate a random error. As a result, there is no correlation between items of the total scale. In tests of the reliability of the measuring tools, the result should be within the range of 0.6–0.8.

A set of terms (in order from the highest to the lowest value of Cronbach's α coefficient) for aspects 2, 3, and 1 obtained acceptable reliability. Aspect 6, which concerned predicting the possibility of implementing CE in a 10- or 20-year perspective, Cronbach's α was 0.462, which can be explained by the fact that the respondents answered "strongly agree" for the 20-year period, whereas for the 10-year period, the vast majority only agreed. The set of aspects 4 and 5 did not receive acceptable reliability. In such a situation, factor analysis was used to indicate the number of factors (sub-aspects) that formed these aspects. Results of the analysis (selection of the optimal number of factors) were based on the screen test and the questions assigned to them were based on the varimax rotational strategy. Four factors were obtained for the set of aspect 4 and two factors for the set of aspect 5.

The values of Cronbach's α coefficients for the subcategories obtained as a result of factor analysis are presented below.

Table 3. Internal consistency of the survey questionnaire.

Analyzed Scope of the Survey Questionnaire	Cronbach's α Coefficient
All survey questionnaire	0.837
Set of terms of aspect 1	0.641
Set of terms of aspect 2	0.727
Set of terms of aspect 3	0.677
Set of terms of aspect 4	0.322
Set of terms of aspect 5	0.193
Set of terms of aspect 6	0.462

As many as 11 terms of aspect 4 concerned the CE requirements for company management. Their diversity was so large that factor analysis, which was used to separate sub–aspects (factors), indicated three strong factors and one weaker one (the last one) (Table 4).

Table 4. Cronbach's α coefficient results for factor analysis of aspect 4.

Additional Classification of Aspect 4 Terms:	Cronbach's α Coefficient
Set of terms for environmental costs and benefits: p 4_4, p 4_10, p 4_11	0.648
Set of terms concerning the external actions related to the implementation of a multi–product strategy: p 4_2, p 4_7, p 4_9	0.757
Set of terms for creating a purchaser market through the use of value: p 4_3, p 4_6	0.606
Set of terms concerning environmental costs generated in relation to suppliers and customers: p 4_1, p 4_5, p 4_8	0.397

The five terms of aspect 5 concerned the reasons for implementing a CE. The factor analysis, which was used to separate sub-aspects (factors), indicated that it was possible to divide them into two factors that could be associated with the external and internal reasons for implementing CE (Table 5).

Table 5. Results of the Cronbach α coefficient for factor analysis of aspect 5.

Additional Classification of the Aspect 5 Terms:	Cronbach's α Coefficient
Set of terms concerning external causes: p 5_1, p 5_3, p 5_4	0.658
Set of terms for internal reasons: p 5_2, p 5_5	0.744

3. Results

The research undertaken has enabled

- the assessment of the attitude of coal power sector companies towards the CE concept;
- the identification of areas stimulating the implementation of CE, allowing one to determine actions for its further development;
- the use of questionnaire using a five-step Likert scale (where 5 was equivalent to "strongly agree" and 1 was assigned to "strongly disagree").

3.1. Aspect 1

The set of terms under aspect 1 concerned the evaluation of the benefits/effects related to implementing CE in a coal power company (Figure 2).

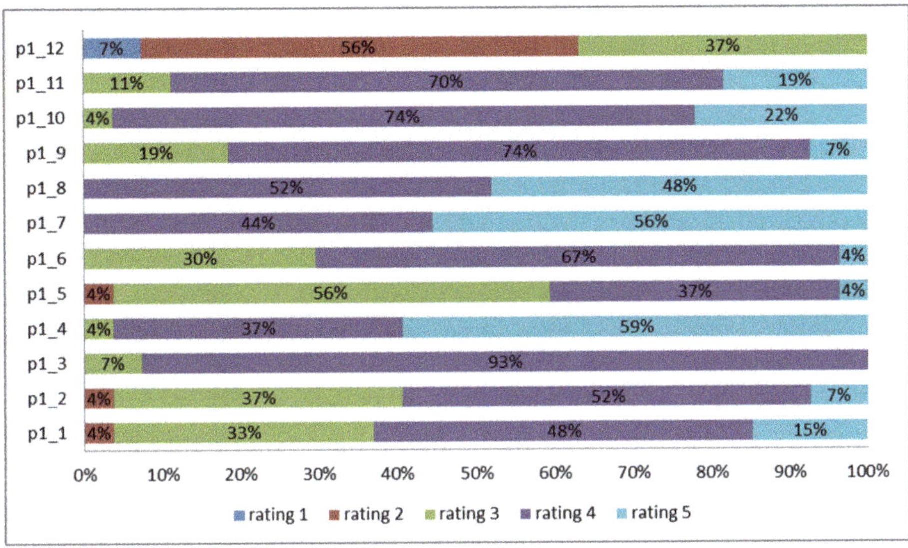

Figure 2. Assessment of the benefits (effects) resulting from implementing CE in coal power engineering.

As regards the assessment of the benefits of implementing CE in the coal power industry, respondents assessed such effects as the increase in efficiency of the use of resources (p1_1), extended product life cycle (p1_2) and establishing relations generating less environmental pressure with consumers (p1_6) in a similar way. In relation to these terms, more than 59% of the respondents answered in the affirmative, no more than 37% of the respondents expressed doubts, while the number of negative opinions did not exceed 4% of the answers obtained. More than 80% of the respondents assessed terms such as introducing more sustainable production (p1_3), reduction of the environmental footprint (p1_4), strengthening standards of environmental management (p1_9), an increase in ecological activity (p1_10) and closing CCP landfills (p1_11). The above-mentioned terms did not obtain negative assessments. The unanimous affirmation (positive assessment) of all respondents was shown in relation to the term generation of products instead of waste (p1_7) and improvement in environmental quality (p1_8). The greatest reflection among the respondents was given to the assessment of the benefits associated with the possibility of establishing less energy-intensive relations with suppliers (p1_5) expressed by 55% of the respondents. The remaining 40% of the respondents gave this term a positive assessment and only 4% a negative one. The statement that received the highest number of negative votes (63%) and no positive assessments was the one concerning benefit resulting from the possibility of reducing the environmental costs of production (p1_12).

3.2. Aspect 2

The set of terms under aspect 2 concerned the assessment of the risks/challenges related to implementing CE in the coal power industry (Figure 3).

As regards the assessment of the risks/challenges associated with implementing CE in the coal power industry, more than 50% of the respondents negatively assessed the risk of reducing a company's competitiveness (p2_4). Difficulty in making the assessment relating to the increase in production costs (p2_1) was expressed by 48% of the respondents, and the need to implement costly and risky investments (p2_2) was expressed by 63% of the respondents.

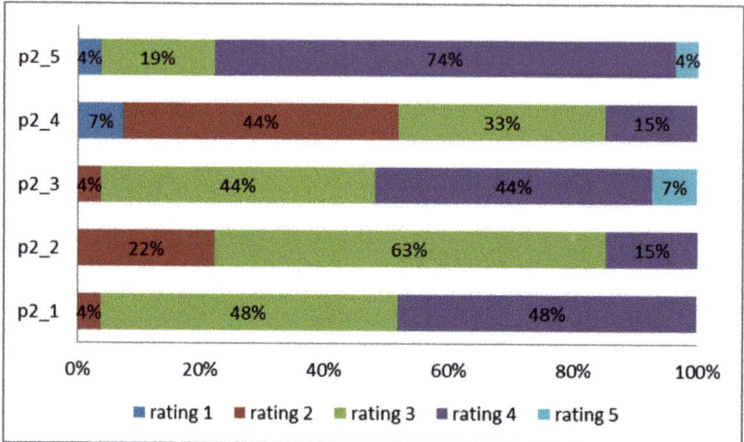

Figure 3. Assessment of the risks/challenges associated with implementing CE in the coal power industry.

The risk that may be posed by implementing CE in the form of an increase in production costs (p2_1) received a greater number of affirmative (48%) than negative (4%) opinions. However, as far as the necessity to carry out investments (p2_2) is concerned, the opposite situation occurred as it received more negative (22%) than positive (15%) opinions. More than half (52%) of the respondents see a challenge in the need to find a way to change the current business model of the process (p2_3), but a large proportion of them (44%) had doubts in assessing this term. The vast majority (78%) of respondents see a risk/challenge in the necessity to go beyond the "plant gates" (p2_5). In total, 18% of the respondents expressed doubts when making the assessment in relation to this term.

3.3. Aspect 3

The set of terms under aspect 3 concerned the assessment of the barriers/obstacles to the implementation of CE in the coal power industry (Figure 4).

In terms of barriers/obstacles that respondents see in the context of implementing CE, over 40% do not agree that raw materials recovered from waste do not find buyers (p3_3). Moreover, in relation to this term, the highest number of opinions (56%) was given to the assessment "neither agree nor disagree." About 50% of the respondents see problems with the lack of implementation of a multi–product strategy of power engineering (p3_1) and the lack of promotion of CCPs as fully fledged products from waste (p3_2). In relation to these two statements, an equally large number of respondents (over 44%) had difficulties in their assessment. More than 70% of respondents admitted that an obstacle to the implementation of CE is the lack of indicators and guidelines for implementing and measuring CE (p3_4). Nearly 90% of the respondents indicated that an obstacle to implementing CE is the fact that the current CE-compliant activities are not appreciated and evaluated (p3_5); moreover, this term was the only one that did not receive a negative assessment.

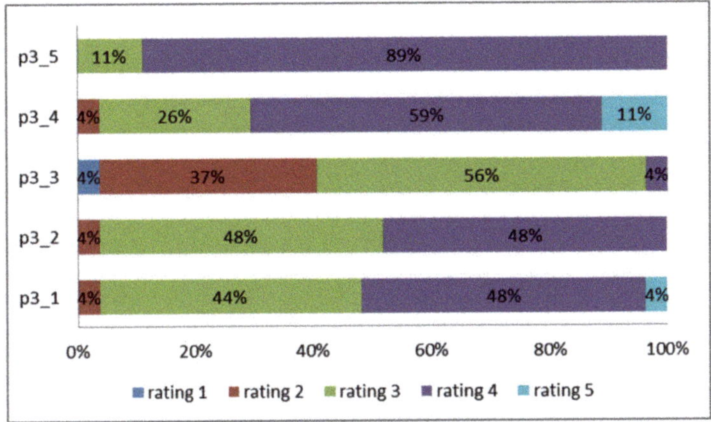

Figure 4. Assessment of barriers/ obstacles to the implementation of CE in the coal power industry.

3.4. Aspect 4

The set of terms under aspect 4 concerned the assessment of CE requirements in relation to the management of a coal power plant (Figure 5).

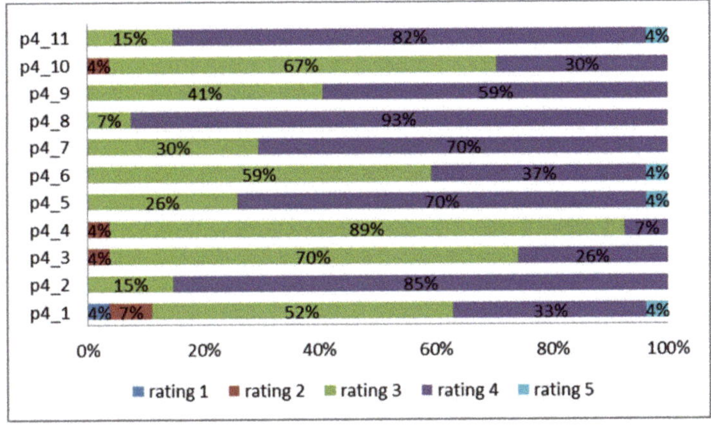

Figure 5. Assessment of the CE requirements in relation to the management of a coal power plant.

The analysis of the assessment of terms related to management activities indicated that more than 50% of the respondents had difficulty in assessing the need to go beyond the boundaries of the plant with responsibility for environmental protection (p4_1) (52%), constructing an income generation logic based on the identification of the value chain (p4_3) (70%), internalization of external environmental costs (p4_4) (89%), creating a buyers–eco-consumers market (p4_6) (59%), and identification of the value chain to specify quantifiable environmental benefits (p4_10) (67%). The number of other responses expressed for these terms was always higher on the positive (affirmative) side.

None of the six terms described below received a negative assessment. These terms include cooperation within clusters (p4_2) (85%), assessment of environmental policy in relation to suppliers and customers (p4_5) (74%), implementation of a multi-product power engineering strategy (p4_7) (70%), recognition of factors representing environmental costs in the expenditure structure (p4_8) (93%), influencing the awareness and actions of other bodies (p4_9) (59%), and integration of environmental policy with a development strategy (p4_11) (85%). The term about which most respondents (41%) were doubtful when expressing their opinion was influencing the awareness and actions of other bodies (p4_9).

3.5. Aspect 5

The set of terms under aspect 5 concerned the assessment of the reasons for the coal power industry to implement CE (Figure 6).

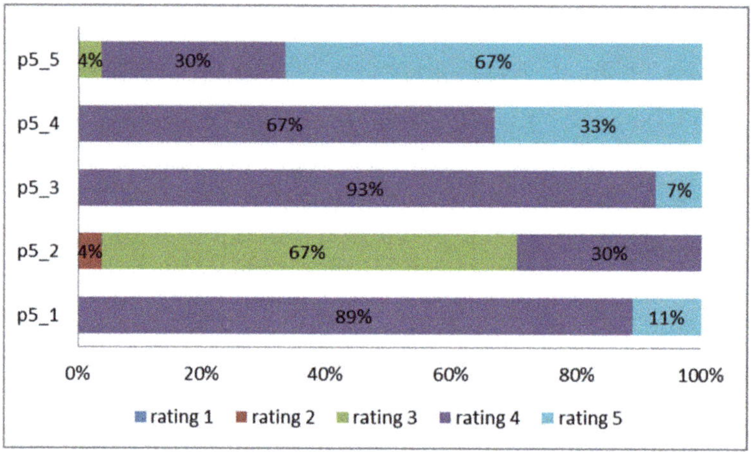

Figure 6. Assessment of reasons for the coal power industry to implement CE.

Respondents had an unequivocally positive assessment of four out of five reasons for implementing CE by coal power companies. These terms included awareness of the importance of environmental protection (p5_1), social benefits (p5_3), the possibility of reducing the product's environmental footprint (p5_4), and economic benefits (p5_5) received 100% of affirmative responses. Only in relation to the term expressing the possibility of reducing costs related to waste management (p5_2) were 4% of respondents doubtful in its assessment.

3.6. Aspect 6

The set of terms under aspect 6 concerned the assessment of the prospect for implementing a CE in the coal power industry (Figure 7).

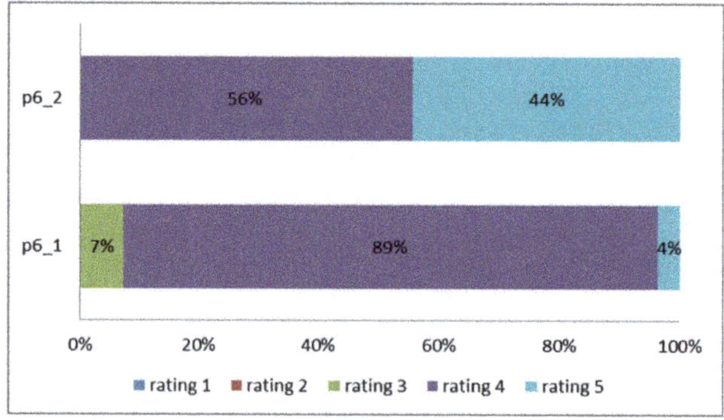

Figure 7. Assessment of the prospect of implementing a CE in the coal power industry.

Concerning the term relating to the possibility of implementing CE in the power industry within 10 or 20 years, the respondents agreed that, in their opinion, there is such a possibility.

4. Discussion

The analysis of the assessments obtained from the respondents for aspect 1 shows that the vast majority of the respondents see the potential of environmental effects that can be achieved as a result of CE. This means that employees are aware of the potential that the implementation of this new trend may bring to the company. The reflection of the respondents shown in relation to the supply chain (benefits from establishing new relations with suppliers) may be caused by the fact that the power industry currently takes into account the quality aspect of the emerging CCPs in specifications for the purchase of raw materials (concerning the requirements for the calorific value of coal, content of ash and sulphur, etc.). Despite the positive assessments expressed in relation to most of the environmental effects generated by the implementation of CE, they do not refer to the initial stage, which is (immediately after the product design process) the stage of material procurement. The supply chain generated is of key importance and is translated into other processes and the environmental costs they create. The supply chain, through transferring the effects of the environmental impacts generated, influences the possibilities of CCP management, in relation to which the culmination of all effects generated in the preceding processes takes place. Therefore, in the opinion of the authors of this study, it can be assumed that in order to implement the CE model it is necessary to identify all links in the value chain and express them in the form of measurable indicators, which will allow the effects produced to be measured. On the other hand, the respondents fail to see that the benefits resulting from the reduction of environmental costs of production by eliminating the generation of waste may be related both to the underestimation of waste management costs (the costs of transport, storage and disposal of waste) and to the lack of internalization of external environmental costs. As a result, this may lead the respondents to believe that current relations do not require additional modifications unless they are additionally imposed by legal requirements.

The analysis of the assessments made for aspect 2 shows that they do not perceive CE as a threat related to disturbance of the position of enterprises on the market; this is due to the fact that they focus primarily on the process of energy generation. On the other hand, CCPs are perceived as a possible secondary product that may bring potential profits–but this is not the direction of the strategies currently being adopted. In the opinion of the respondents, the most important thing for enterprises is that the carrying out of a CE brings no increase in production costs. This aspect can be directly combined with the above described negative assessment expressed in relation to the benefits of implementing CE in the form of reducing environmental production costs through eliminating waste generation. This translates into further consequences, which are located in the concerns about the need to find a way to change the business model and the need to go beyond the plant gates related, for example, to the carrying out of a life cycle assessment. This may result from the lack of a consistent approach to environmental protection throughout the entire value chain generated by the company, which means that it is limited to analyses within the plant gates.

The answers obtained for aspect 3 prove that the respondents see the potential and the need to implement a multi–product power engineering strategy in the context of the possibilities that CCP management brings. However, at the same time, the potential for their sale is not utilized. The respondents also show that in their opinion activities carried out so far are CE compatible. However, the lack of guidelines for CE measurement is an obstacle in verifying this opinion.

As far as aspect 4 is concerned, the answers received confirm that requirements set by the CE, connected with the necessity to go beyond the plant gates and to analyze the impact of the activity throughout the whole life cycle determining the areas where the activities enabling the implementation of CE should be commenced. All activities where it is necessary to go beyond the plant gates and their measurement are a big barrier (e.g., introducing economic symbiosis), as are connecting with data availability and modelling.

The answers obtained for aspect 5 indicate that four out of the five reasons for implementing CE are important for coal power companies. It should be emphasized that generating economic profits was the only positively evaluated aspect for which the majority of respondents indicated the answer "I strongly agree." The respondents' doubts as to the possibility of reducing the costs of waste management may be explained by the fact that, at present, they are unable to direct their thinking towards the goal of eliminating the generation of waste in order to generate products/goods. This may be caused by focusing only on generating profits for the company from the leading activity, which is energy production (and focusing on current profits).

The answers obtained for aspect 6 indicate that the respondents envisage implementing CE in the next decade. This may be due to the decisions of the company that, through its declared environmental policy and the resulting pro-environmental attitude, will take actions related to implementing CE spontaneously. There is also an option that CE implementation will result from the need to adapt to anticipated changes in legislation.

In contemporary Polish strategic document and policy, i.e., National Energy Policy until 2040, most changes focus on the reduction emissions of CO_2 and other air pollutant from burning coal and less on circularity and waste management. This is a global important priority, but despite the efforts of many countries in their policies and strategies to reduce energy production from fossil fuels, a significant share of the world's energy supply, as well as the CO_2 emissions, is still generated from coal.

Global CO_2 emissions from fossil fuels combustion and processes further increased by 1.9% in 2018 compared to the previous year reaching a total of 37.9 Gt CO_2. More detailed data about anthropogenic greenhouse gas emissions can be also found in Emissions Database for Global Atmospheric Research (EDGAR). Since 2015, the EU–28 share of global fossil CO_2 emissions has remained constant at 9.6% which is the equivalent of seven tonnes of CO_2/cap/yr. The largest contributor in 2017 was Germany with a 22.4% share, followed by the United Kingdom (10.7%), Italy (10.2%), France (9.5%), and Poland (9%) [23]. Poland, however, noted a significant decrease (29%) in CO_2 emissions from the power industry in the years 1990–2017. In Poland, the emissions reached 8.3 tonnes of CO_2/cap/yr in 2017, whereas in Germany, they reached 9.7 tonnes of CO_2/cap/yr [24].

Wider and full utilization of CCPs in Poland can also have a beneficial impact on the reduction of energy use and CO_2 emissions in the industry. Beneficial use of CCPs as raw materials in manufacturing construction materials is well established both in literature [25] and the practice of many EU-15 countries as well as Japan, Korea, and China, where the utilization rate has reached over 70%. CPPs are mostly used for cement and concrete, road construction as well as for geotechnical and filling applications, including reclamation, as CCPs have established product standards or technical guidelines. There are, however, proposed solutions, such as

1. Extraction of aluminum content from coal fly ash [26] in Indian thermal power plants, which generate around 1 million tonnes with around 25–35% Al_2O_3 and 50–60% SiO_2. Al was extracted by sulphuric acid leaching with calcium sulphate as leach residue. At the conclusion of this process, the leachate is suitable for downstream operations to precipitate alumina.
2. Lithium recovery [27] from coal fly ash by a combination of an intensified acid leaching process and pre–desilication.
3. A method for the extraction from coal ash of iron, Al, and Ti (United States Patent 4567026) [28].
4. Research on the recovery of trace and rare metals from coal and the products of their traditional and thermal enrichment in Poland. Better value raw materials for the metallurgy industry are obtained through a careful selection of coals and the use of carefully selected methods of enriching metal concentrates. This also ensures a significant economic impact on the cost of production [29].
5. Furthermore, Hycnar and Tora [29] note research carried out in Poland that allows one to set up the technology for producing metal concentrates such as germanium and gallium oxide concentrate, iron oxides, aluminum oxide, and calcium oxide.

6. Analysis of the available technologies, the opportunities for their improvement, and the sources of financing in connection with the implementation of CE principles, as well as the results of research on assessing the implementation of CE in large energy concerns indicate that, along with the development of CE policy in Poland, there is a large potential for the full utilization of CCPs.

5. Conclusions

From the assessment of the respondents and legal changes in EU and new energy strategy Poland, it can be concluded that the implementation of CE requires unanimity in actions and ensuring their measurability, which can be achieved by the introduction of the circular business model and its indicators.

Based on the terms of all analyzed aspects, it can be concluded that

- The respondents do not understand the context of implementing CE in the value chain (aspect 1—benefits/effects of CE implementation);
- The solution to the problem of concerns about the need to find a way to change the business model toward CE by assessing the entire value chain, without limiting it to the so-called "plant gates." Such changes can create the increase of production cost (aspect 2—risks/challenges related to CE implementation);
- The biggest obstacle to CE implementation is the lack of guidelines for CE implementation and measurement. Companies need indicators to be able to assess current activities that, in the opinion of respondents, are consistent with the CE (aspect 3—barriers related to CE implementation);
- CE requirements determine areas of necessary modifications to the current way the company is managed. In relation to these areas, it is advisable to propose measurement indicators so that companies can verify the implementation of CE and indicate actions for further modifications. Moreover, it can also be concluded that the respondents, although they show the need for indicators, are not prepared to measure them in the value chain (aspect 4—CE requirements in relation to the management of a company);
- Companies are willing to implement CE due to the economic, environmental, and social aspects. However, in order to capture the economic effects, it is necessary to present the environmental effects in the form of measurable indicators, showing the value that can be obtained when generating a product instead of waste (aspect 5—reasons for CE implementation);
- The respondents see the need to prepare for the changes that are expected in the current business model. The answer to this need is the necessity to develop a circular business model (together with indicators for its measurement), the implementation of which will enable the necessary changes to be made (aspect 6—prospects of implementing CE).

In relation to the results obtained, it can be concluded that the respondents see potential in the modification of the current method of company management in the context of the CCPs generated, but they lack precise guidelines on how to do it (areas of action and indicators), which determines the direction and need for further research in this area. Successful implementation of the circular economy rules in Poland creates conditions for the promotion of zero waste technologies. Polish companies could analyze in more detail and assess good practices from different countries and focus on closer cooperation with scientific institution. It additionally allows them to apply for financial support for such eco-friendly investments.

Author Contributions: Conceptualization, J.K. and A.B.; methodology J.K.; validation, A.B. and J.K.; formal analysis, A.B.; investigation, A.B. and J.K.; resources A.B.; data curation A.B. and J.K.; writing—original draft preparation, A.B.; writing—review and editing, J.K.; visualization, A.B. and J.K.; supervision, J.K.; funding acquisition, J.K. All authors have read and agreed to the published version of the manuscript.

Funding: The publishing of this work was supported by National Center for Research and Development as a part of the oto–GOZ project and AGH University of Science and Technology in Cracow, Poland.

Conflicts of Interest: The authors declare no conflict of interest.

References

1. Harris, D.; Heidrich, C.; Feuerborn, J. Global Aspects on Coal Combustion Products. Available online: https://www.coaltrans.com/insights/article/global-aspects-on-coal-combustion-products (accessed on 15 December 2019).
2. Yao, Z.T.; Ji, X.S.; Sarker, P.K.; Tang, J.H.; Ge, L.Q.; Xia, M.S.; Xi, Y.Q. A comprehensive review on the applications of coal fly ash. *Earth-Sci. Rev.* **2015**, *141*, 105–121. [CrossRef]
3. Seshadri, B.; Bolan, N.S.; Naidu, R.; Wang, H.; Sajwan, K. Clean Coal Technology Combustion Products. *Adv. Agron.* **2013**, *119*, 309–370.
4. Ash Development Association of Australia, Coal Combustion Utilisation. Available online: http://www.adaa.asn.au/resource-utilisation/ccp-utilisation (accessed on 15 December 2019).
5. Homepage of World Wide Coal Combustion Products Network. Available online: http://www.wwccpn.org/ (accessed on 19 November 2018).
6. Xing, Y.; Guo, F.; Xu, M.; Gui, X.; Li, H.; Li, G.; Xia, Y.; Han, H. Separation of unburned carbon from coal fly ash: A review. *Powder Technol.* **2019**, *353*, 372–384. [CrossRef]
7. Park, J.Y.; Chertow, M.R. Establishing and testing the "reuse potential" indicator for managing wastes as resources. *J. Environ. Manag.* **2014**, *137*, 45–53. [CrossRef] [PubMed]
8. Szczygielski, T.; Niewiadomski, M. Zarys modelu biznesowego dla energetyki. Bezodpadowa Energetyka Węglowa (BEW) (Outline of a business model for zero waste coal power (ZWCP)), Materials from the meeting of the Circular Economy in Poland Working Group, 25 May 2017, Warsaw.
9. Kulczycka, J.; Dziobek, E.; Szmiłyk, A. Challenges in the management of data on extractive waste—The Polish case. *Miner Econ.* **2019**. [CrossRef]
10. Millward-Hopkins, J.; Purnell, P. Circulating blame in the circular economy: The case of wood-waste biofuels and coal ash. *Energy Policy* **2019**, *129*, 168–172. [CrossRef]
11. Ministerstwo Aktywów Państwowych. Updated Draft Version of Energy Policy of Poland until 2040. Available online: https://www.gov.pl/web/aktywa-panstwowe/polityka-energetyczna-polski-do-2040-r-zapraszamy-do-konsultacji (accessed on 23 November 2018).
12. Krajowy Plan Na Rzecz Energii i Klimatu Na Lata 2021–2030. Available online: https://www.gov.pl/web/klimat/krajowy-plan-na-rzecz-energii-i-klimatu (accessed on 20 February 2020).
13. EC-European Commission. Directive 2010/75/EU of the European Parliament and of the Council of 24 November 2010 on industrial emissions. *Off. J. Eur. Union L* **2010**, *334*, 2010.
14. Jarema-Suchorowska, S.; Zupa-Marczuk, P.; Głowacki, E. BAT/BREF—Review of LCA requirements for emissions of pollutants and their impact on CCPs economy. In Proceedings of the 26th International Conference "Ashes from Power Generation", Sopot, Poland, 8–10 October 2019.
15. Kulczyka, J.; Dziobek, E.; Ubeman, R. Industrial Symbiosis for the Circular Economy Implementation in the Raw Materials Sector—The Polish Case. In *Industrial Symbiosis for the Circular Economy: Operational Experiences, Best Practices and Obstacles to A Collaborative Business Approach*; Salomone, R., Cecchin, A., Deutz, P., Raggi, A., Cutaia, L., Eds.; Springer: Berlin, Germany, 2020; pp. 73–86. Available online: https://www.springer.com/gp/book/9783030366599 (accessed on 20 May 2020).
16. Ahmaruzzaman, M.A. Review on the utilization of fly ash. *Prog. Energy Combust. Sci.* **2010**, *36*, 327–363. [CrossRef]
17. Blissett, R.; Rowson, N.A. Review of the Multi-Component Utilisation of Coal Fly Ash. *Fuel* **2012**, *97*, 1–23. [CrossRef]
18. Huang, T.Y.; Chiueh, P.T.; Lo, S.L. Life-cycle environmental and cost impacts of reusing fly ash. *Resour. Conserv. Recycl.* **2017**, *123*, 255–260. [CrossRef]
19. Izquierdo, M.; Querol, X. Leaching behaviour of elements from coal combustion fly ash: An overview. *Int. J. Coal Geol.* **2012**, *94*, 54–66. [CrossRef]
20. Kristensen, H.S.; Mosgaard, M.A. A review of micro level indicators for a circular economy – moving away from the three dimensions of sustainability? *J. Clean. Prod.* **2019**, *243*, 118531. [CrossRef]
21. Krajowa Inteligentna Specjalizacja, Circular Economy—Water, Fossil Raw Materials, Waste. Available online: https://smart.gov.pl/en/circular-economy-water-fossil-raw-materials-waste (accessed on 20 November 2019).
22. Gonenc, H.; Scholtens, B. Environmental and Financial Performance of Fossil Fuel Firms: A Closer Inspection of their Interaction. *Ecol. Econ.* **2017**, *132*, 307–328. [CrossRef]

23. Crippa, M.; Oreggioni, G.; Guizzardi, D.; Muntean, M.; Schaaf, E.; Lo Vullo, E.; Solazzo, E.; Monforti-Ferrario, F.; Olivier, J.G.J.; Vignati, E. *Fossil CO$_2$ and GHG Emissions of All World Countries—2019 Report*; EUR 29849 EN; Publications Office of the European Union: Luxembourg, 2019; ISBN 978-92-76-11100-9. [CrossRef]

24. Lelek, Ł.; Kulczycka, J. Life Cycle Modelling of the impact of coal quality on emissions from energy generation. *Energies* **2020**, *13*, 1515. [CrossRef]

25. Association of State and Territorial Solid Waste Management Officials. Beneficial Use of Coal Combustion Residuals—Survey Report, September 2012. Available online: http://astswmo.org/files/policies/Materials_Management/2012-09-BUTF-BU_of_CCRs_Report.pdf (accessed on 13 January 2020).

26. Tanvar, H.; Chauhan, S.; Dhawan, N. Extraction of aluminum values from fly ash. *Mater. Today* **2018**, *5*, 17055–17063. [CrossRef]

27. Li, S.; Qin, S.; Kang, L.; Liu, J.; Wang, J.; Li, Y. An efficient approach for lithium and aluminum recovery from coal fly ash by pre-desilication and intensified acid leaching processes. *Metals* **2017**, *7*, 272. [CrossRef]

28. Method for Extraction of Iron Aluminum and Titanium from Coal Ash. Available online: http://www.freepatentsonline.com/4567026.html (accessed on 13 January 2020).

29. Hycnar, J.; Tora, B. Analiza zawartości wybranych metali w węglach i produktach ich spalania. *CUPRUM–Czasopismo Naukowo-Techniczne Górnictwa Rud.* **2015**, *2*, 157–168.

Article

Numerical Modelling and Experimental Verification of the Low-Emission Biomass Combustion Process in a Domestic Boiler with Flue Gas Flow around the Combustion Chamber

Przemysław Motyl [1],* , Danuta Król [2], Sławomir Poskrobko [3] and Marek Juszczak [4]

[1] Faculty of Mechanical Engineering, University of Technology and Humanities in Radom, 26-600 Radom, Poland

[2] Faculty of Energy and Environmental Engineering, Silesian University of Technology, 14-100 Gliwice, Poland; dankrol@wp.pl

[3] Faculty of Civil Engineering and Environmental Sciences, Białystok University of Technology, 15-351 Białystok, Poland; drposkrobko@wp.pl

[4] Faculty of Environmental Engineering and Energy, Poznan University of Technology, 60-965 Poznań, Poland; marekjuszczak8@wp.pl

* Correspondence: p.motyl@uthrad.pl

Received: 8 October 2020; Accepted: 31 October 2020; Published: 9 November 2020

Abstract: The paper presents the results of numerical and experimental studies aimed at developing a new design of a 10 kW low-emission heating boiler fired with wood pellets. The boiler is to meet stringent requirements in terms of efficiency ($\eta > 90\%$) and emissions per 10% O_2: $CO < 500$ mg/Nm^3, $NOx \leq 200$ mg/Nm^3, and dust ≤ 20 mg/Nm^3; these emission restrictions are as prescribed in the applicable ECODESIGN Directive in the European Union countries. An innovative aspect of the boiler structure (not yet present in domestic boilers) is the circular flow of exhaust gases around the centrally placed combustion chamber. The use of such a solution ensures high-efficiency, low-emission combustion and meeting the requirements of ECODESIGN. The results of the numerical calculations were verified and confirmed experimentally, obtaining average emission values of the limited gases $CO = 91$ mg/Nm^3, and $NOx = 197$ mg/Nm^3. The temperature measured in the furnace is 450–500 °C and in the flue it was 157–197 °C. The determined boiler efficiency was 92%. Numerical calculations were made with the use of an advanced CFD (Computational Fluid Dynamics) workshop in the form of the Ansys programming and a computing environment with the dominant participation of the Fluent module. It was shown that the results obtained in both experiments are sufficiently convergent.

Keywords: biomass combustion; pellet boiler; CFD modeling; renewable heating

1. Introduction

Excessive CO_2 emissions to the atmosphere, the threat of smog (especially in the vicinity of large urban agglomerations), and appropriate legal regulations force the low energy industry to eliminate the use of fossil solid fuels. In the European Union countries, hard coal has almost been eliminated from the market as a fuel intended for small prosumer energy. Solid renewable fuels, wood biomass, and agro biomass remained on the market. Combustion of these fuels in low-power (domestic) boilers can undoubtedly contribute to reducing CO_2 emissions to the atmosphere. However, this will not make onerous and dangerous smog disappear [1–3]. To meet the care of the natural environment, the European Union Parliament imposed the obligation to conduct the biomass combustion process for low-power boilers intended for heating small objects (including residential buildings), so that the emission of combustion products to the atmosphere meets the requirements of the ECODESIGN standard in the scope of CO < 500 mg/Nm^3, NOx < 200 mg/Nm^3, and dust < 40 mg/Nm^3 for O_2 = 10 mg/Nm^3 in the flue gas [4–6]. This standard is effective in European Union countries from 2020. The requirements of the standard (when it comes to low-power boilers fired with biomass) mainly concern the limited values of CO and NO_x emissions and the conduct of the combustion process with high energy efficiency—the process high-efficiency combustion is associated with the low-emission process. Meeting the abovementioned requirements obliges boiler manufacturers (especially in countries where solid fuel boilers are widely used) to take actions aimed at continuous improvement of their products, designing a range of products adapted to legal requirements. Considering the needs of atmosphere protection and the related market requirements, in many research units actions are taken to identify biomass fuels burned in low-power boilers, model their combustion process, and increase fuel-burning efficiency to meet the standards in force in the European countries (EC). Given the stringent emission standards, the considerable potential of biomass in EC countries encourages the development of biomass combustion technology in low-power boilers [7–10]. Therefore, it is necessary to support the activities discussed in [11], where the authors justify the social and economic need for effective use of this fuel in the small domestic energy industry. Such actions are justified by the sustainable development policy implemented in the European Union countries. In [12], the efficiency of burning wood pellets in domestic boilers, with different loads and different configurations of burner settings was discussed and compared. The authors stated that high emission and energy efficiency depend (when it comes to structural changes) on the configuration of the burner—higher efficiency is obtained at the lower power supply. In the paper [13], the authors analysed the work of several hundred boilers fired with biomass pellets. These boilers were installed in residential buildings in Germany, Austria, Great Britain, and Spain. Despite the interesting operational data, it is worth noting that the monitored boilers were characterized by a typical design of the boiler furnace and convective part. The burner was located centrally at the bottom of the spacious furnace. The convective part was located behind the hearth, in a system of single-draft, self-cleaning ducts with a circular cross-section (flame tubes). The exhaust gas flowed through these channels to the flue. The furnace and convection part were separated from the surroundings with a water jacket and a layer of insulation. Such boilers (as indicated in the data contained in [13]) are widely used in EC countries. Other—relatively popular—boiler structures are presented in [14]. The research was aimed at obtaining useful data (for two different commercial low-power boilers), which concerned the intensity of sediment formation on the walls of heat exchangers. The first boiler is a two-draft boiler with a water jacket in the hearth and in the convection part, which is the second flue gas draft. In the upper part of the furnace and convective part—in order to increase the heat exchange surface—there were tufts of water tubes flushed by exhaust gases. The exhaust fumes from the convection part located behind the combustion chamber were directed to the smoke duct. The second boiler is a common one-draft structure with a flue gas/water heat exchanger located directly above the tube burner. The authors noted that the ability to settle solid particles was lower in the case of a boiler with a vertical single-tube heat exchanger. The reason for the appearance of smaller deposits in a single-pipe boiler is the stronger turbulence of flue gas flow through the pipes in which screw turbulators were additionally installed. Similar studies are presented in [15]. The researchers' interests were tubular heat exchanger boilers. They examined the impact of installing additional elements of the tubes in the heat exchanger. Based on the results, they found that an increase in the number of tubes (i.e., an increase in the heat exchange surface) has an impact on the increase in boiler efficiency, with an increase in the CO content in the flue gas. There were no changes in solid particle emissions. In the paper [16], the possibilities of modernizing the heat-flow system of a two-draft boiler with a water jacket in the furnace and in the convection part, constituting the second flue gas, were pointed out. The added fan increased the turbulence of the exhaust gas flow in the upper part of the furnace and convection chamber.

Observations of the test results based on the above review of biomass combustion technology in small boilers show that the boiler designs are based mainly on the concept of hot flue gas flow in multi-pipe (usually two or three-pipe) channels with a water jacket. The channels are arranged according to the following scheme: the hearth, behind the hearth, successive convection channels (one behind the other) or the hearth, and other convection channels above the hearth (one above the other). These can be rectangular ducts or multi-tube flue gas ducts (smoke tubes). Tubes (smoke tubes) are also often used in the construction of flue gas/water heat exchangers with single-line flue gas flow, where the tube exchanger is located above the furnace.

The aim of the work was to build a prototype of a 10 kW boiler with an innovative furnace design and convection flue gas circulation channels. The boiler is designed to burn pellets made of wood biomass in the continuous feeding mode. The aim of the work was achieved by carrying out numerical and experimental research. Numerical tests included the following: (i) modelling the height of the combustion chamber by analysing the combustion process of syngas produced from biomass in the retort burner; (ii) flue gas flow, temperature distribution and heat exchange calculation in the furnace and convection channels; and, (iii) concentration of gaseous products of CO, CO_2 combustion: NO_x and O_2. The experimental tests consisted in the verification of the obtained results from the numerical analysis in the scope of the following: (i) temperature measurement in the furnace and in the flue, (ii) measurement of gas concentration of combustion products, and (iii) determination of the boiler efficiency. The key assumption of the innovative design concept was to place the furnace with the burner in the central part of the boiler body. The adopted concept of the location of the furnace forces the circulation of hot exhaust gases in a four-pass flow. Flow I is the hearth, Flows II, III, and IV form a system of circular convection flow of exhaust gases around the hearth. Numerical calculations and then the verification of compliance of the calculation results with the results of emission measurements play an important role in the process of design development. The offered boiler is the only solution of this type with a characteristic central location of the furnace and circulating flue gas.

2. Materials and Methods

The paper proposes a 3-stage, semi-empirical procedure for creating a numerical model of the boiler (Figure 1). After determining the target boiler power and combustion technique in the burner, an experiment was carried out to determine the composition of the synthesis gas, which is the product of gasification processes taking place in the burner. These data were used as the boundary condition for the burner model in the cylindrical test chamber. Based on the information (combustion chamber height, temperature distribution, CO and O_2 concentrations) adopted from calculations of syngas combustion in the cylindrical area (Figure 2), a numerical model of the actual boiler combustion chamber was made. Multi-variant simulations were carried out to suggest optimal chamber geometry. The last stage was the prototype of the boiler and the verification of model assumptions.

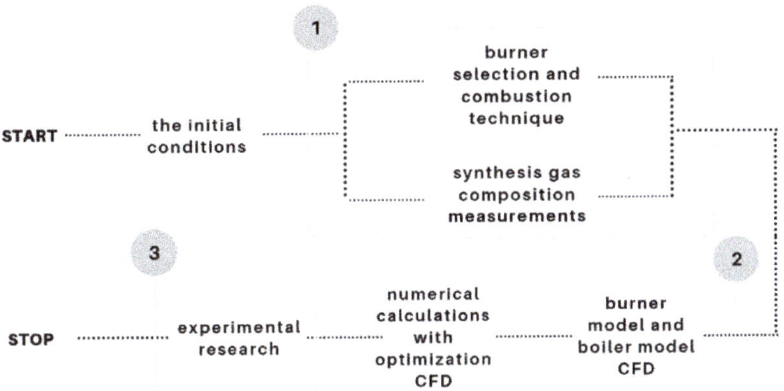

Figure 1. Scheme of 3-stage, semi-empirical modelling of the boiler.

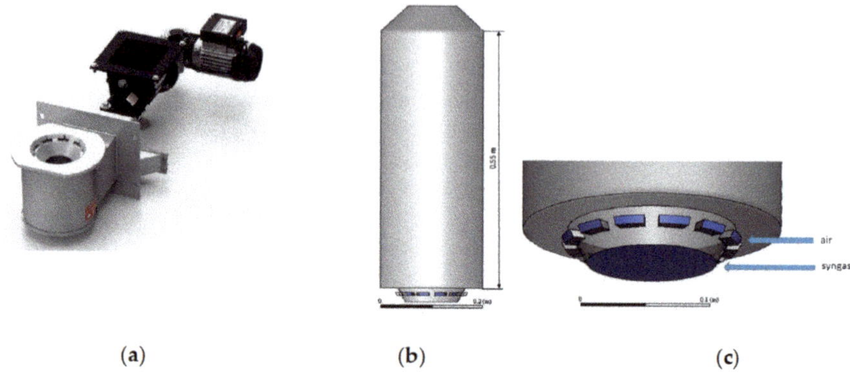

(a)   (b)   (c)

Figure 2. Retort burner for gasification of pellets: (**a**) view of the burner structure, (**b**) burner model in a cylindrical chamber, (**c**) burner model.

2.1. Composition and Calorific Value of the Generator Gas

In the boiler furnace, the generator gas (syngas) will be burned, which was obtained in the process of wood biomass gasification; the wood biomass was in the form of pellets with a diameter of 3 mm and a length of about 10–15 mm (average values). The fuel characteristics of the tested biomass are shown in Table 1. Samples for analytical determinations were taken in accordance with the PN-EN 14778:2011 standard Solid biofuels–Sampling. Fuel properties were determined according to PN-EN 14774-1:2010 Solid biofuels—Determination of moisture content—Drying method—Part 1: Total moisture; PN-EN 14775:2010 Solid biofuels—Determination of Ash; PN-EN 14918: 2010 Solid biofuels—Determination of calorific value; PN-EN 15104:2011 Solid biofuels—Determination of total carbon, hydrogen and nitrogen content—Instrumental methods; PN-EN 15289: 2011 Solid biofuels—Determination of total sulphur and chlorine. The gasification process was carried out in the retort burner chamber (Figure 2a) with an added cylindrical chamber (Figure 2). The gasification factor was air. During gasification, the air demand index expressed as a fuel/air ratio F/A was 0.72, and fuel consumption F = 0.000668 g/s, air gasification consumption A = 0.000927 g/s. Syngas was obtained; its composition in the dry state is presented in Table 2. The concentrations of individual components were determined using the GAS 3000 syngas analyser. The calorific value of syngas was calculated based on Formula (1) [17]:

$$LHV = 126[CO] + 108[H_2] + 359\ [CH_4],\tag{1}$$

where [CO], [H$_2$], and [CH$_4$] percentage volumetric shares of syngas flammable components are indicated.

Table 1. Fuel properties of wood pellets.

Fuel	C [%] **	H [%] **	S [%] **	Cl [%] **	N [%] **	O [%] **	Flammable Fraction *	A [%] *	Moisture Content [%]	LHV kJ/kg *	LHV kJ/kg
Wood pellets	48.93	6.48	0.02	0.01	0.93	43.54	99.4	0.6	9.0	18,145	16,165

* Expressed on a dry free basis, ** Expressed on a dry ash free basis.

The method of measuring the synthesis gas composition is shown in Figure 3. In order to separate the syngas from the environment, a cylindrical syngas chamber (1) was added to the burner, to which this gas flows from the furnace (2). Gas samples were collected through the syngas analyser system (9) to measure its composition. The gasification air (primary air) was supplied by the blowing fan (7) and through the ducts arranged on the perimeter of the lower part (4). This air enters the fuel layer fed continuously to the furnace (2). Mineral residue after the gasification process automatically moves to the top of the furnace, from where periodically, after sliding up the chamber (1), is removed outside. After removing ash, the chamber (1) returns to the sealed position and in this position the syngas samples are taken through the measuring system (8) and (9). The upper air blowing ducts (secondary air) during the synthesis gas composition test are plugged. If the syngas chamber (1) for measuring the synthesis gas composition is removed under normal boiler operation conditions, then syngas combustion will take place thanks to the secondary air supplied through the cleared channels (5) around the periphery of the upper part of the furnace (Figures 2 and 3).

Table 2. Syngas components from biomass.

Component	Unit	Value
CO_2	%	15.1
CO	%	18.4
H_2	%	14.2
CH_4	%	3.3
N_2	%	49.0
LHV	kJ/Nm^3	4857

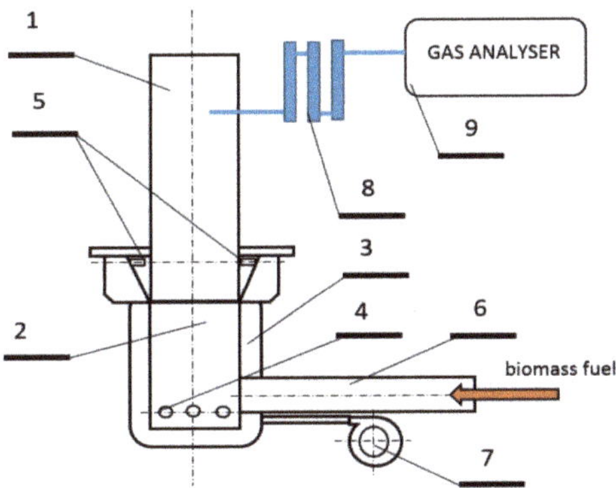

Figure 3. Gasification of biomass fuel in the retort burner: (1)—syngas chamber, (2)—gasification furnace, (3)—air chamber, (4)—gasification air inlets, (5)—combustion air inlets, (6)—screw feeder, (7)—air blower, (8)—syngas purification battery, (9) measurement of syngas composition—GAS 3000 syngas analyser.

2.2. Numerical Modelling of Combustion Processes

Numerical modelling of the construction of a new boiler type began with the analysis of the combustion process in the burner (Figure 2a) of syngas produced in the process of wood pellets gasification. The composition of syngas obtained during the gasification process is provided in Table 2. In Figure 2b, a model of the burner in a cylindrical chamber is presented, which served as a starting point for determining the correct chamber geometry of the designed boiler. The essence of the numerical calculations was the following: (i) determining the height at which, at 100% burner load, all combustible particles undergo combustion in the mixture of synthesis gas and air; and, (ii) determining the temperature distribution that governs the efficiency and low-emission combustion. Numerical calculations were made assuming a syngas flow corresponding to 10 kW with an air excess factor $\lambda = 2$. The composition of syngas corresponded to the experimental (real) data given in Table 2. The air flow was calculated according to the following relationships (2) and (3):

$$O_t = \frac{1}{2}(H_2 + CO) + 2CH_4 - O_2 \left[m^3 O_2 / m^3 \ of \ fuel \right] \tag{2}$$

$$L_t = \frac{O_t}{0.21}\left[\text{m}^3 \text{ of air}/\text{m}^3 \text{ of fuel}\right] \tag{3}$$

In the second step of numerical modelling, a full boiler model with flue gas flow and positioning of the stake according to the original concept was made.

The numerical grid of the cylindrical combustion chamber with the burner was built of 622,000 polyhedral cells, while the numerical grid of the full boiler model consists of 945.000 polyhedral cells. In order to simulate the gas phase reactions within the full geometry of a biomass furnace in CFD, models for turbulence and radiation were selected [18]. The implementation of the realizable k-ε (enhanced wall treatment) turbulence model enables the Reynolds turbulent equations to be closed. The discrete ordinates radiation model was used to simulate the radiation heat transfer. Combustion reactions are complex and simplified into a series of global reactions with CFD modelling. The hybrid Finite-Rate/Eddy-Dissipation model [18] with the typical reactions scheme can be summarized in the form of kinetics rates [19,20]:

$$CH_4 + 1.5O_2 \rightarrow CO + 2H_2O \quad Ar \, (s-1) = 5.012 \times 10^{11} \quad E'r \, (\text{J/kmol}) = 2.0 \times 10^8 \tag{4}$$

$$CO + 0.5O_2 \rightarrow CO_2 \quad Ar \, (s-1) = 2.239 \times 10^{12} \quad E'r \, (\text{J/kmol}) = 1.7 \times 10^8 \tag{5}$$

$$H_2 + 0.5O_2 \rightarrow H_2O \quad Ar \, (s-1) = 9.870 \times 10^8 \quad E'r \, (\text{J/kmol}) = 3.1 \times 10^7 \tag{6}$$

The gas phase absorption coefficients are calculated using the weighted-sum-of-grey gases model (WSGGM). Boundary conditions are as described in Table 3. Water cooling was installed on the wall in the cylindrical chamber model. The necessary settings were fixed according to the documentation for the Ansys Fluent packet. In a similar method of modelling, the operation of a biomass gasifying retort burner is presented in [21].

Table 3. Boiler's full-load data (N—standard temperature and pressure: 20°C, 1013 hPa).

Name	Parameter	Unit	Value
Wall–heat exchanger (walls cooled by water)	Convection coefficient	W/m^{2}K	2500
	Water temperature	K	338
	Steel thickness	mm	5
Inlets	Q of syngas	Nm^3/s	0.002059
	Q of air	Nm^3/s	0.004314
	Syngas temperature	K	1123
	Air temperature	K	300
Outlet	Vacuum	Pa	−15

2.3. Experimental Set up and Procedure

Measurements were carried out in the boiler installation belonging to the Poznan University of Technology, Division of Heating, Air Conditioning and Air Protection. Experiments were performed in almost real-life conditions of small power boilers. Boiler installation with measurement equipment is presented in Figure 4.

Pollutant concentrations in the flue gas and flue gas temperature were measured using the Vario Plus (manufacturer—MRU GmbH Germany) flue gas analyser. CO and C_xH_y concentrations were measured using the infrared procedure. Oxygen (O_2), nitric oxide (NO), nitrogen dioxide (NO_2) concentrations were measured with electrochemical cells. The gas analyser Vario Plus also calculated the air excess ratio and chimney loss for the boiler.

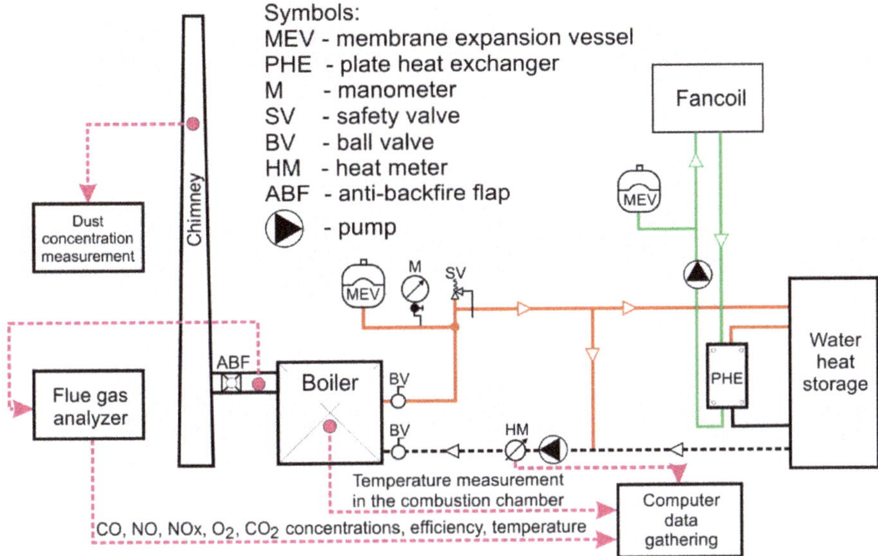

Figure 4. Boiler installation with measurement equipment.

Heat received by the boiler water and boiler heat output were measured with a Kamstrup heat meter. Fuel-mas was measured with Sartorius weight. The boiler heat efficiency was calculated as heat transferred to the boiler water, divided by fuel-mas, multiplied by fuel lower heating value.

3. Results and Discussions

3.1. Results of Numeric Simulations

Examples of the results of temperature distribution, combustion gas products, and exhaust gas paths are given in Figures 5–11. Table 4 presents the temperature and flue gas composition values in the outlet plane to the chimney from the target boiler model. Analysing the results (Figure 5c), it should be noted that only in the central part of the cylindrical chamber are there favourable conditions for the rapid burning of carbon monoxide (the temperature exceeds 882 K, i.e., the flash point of CO and other compounds, i.e., CH_4 and H_2). The observed CO concentration may result in the formation inside the boiler chamber, of low temperature and low O_2 content zones, which are unfavourable for reducing flammable compounds. Consequently, this will lead to their emission through the chimney to the atmosphere. Based on the results of the numerical calculations of syngas combustion (syngas composition Table 2) in the burner (Figure 5), with the assumed geometry of the burner (Figure 6), the height of the combustion chamber of the boiler h = 285 mm was determined. Figure 7 shows the calculated exhaust gas path from the burner through the combustion chamber located in the central part of the boiler. It should be emphasized that, in the proposed geometry of the combustion chamber in the boiler (Figures 6 and 2a), the effect of flue gas swirling was obtained (Figure 7), which promotes afterburning of carbon monoxide and also intensifies heat exchange with the water jacket. The combustion of carbon monoxide is additionally favoured by the high temperature zone, which exceeds the flash point of CO of 973 K (700 °C) in the cross section (Figures 8 and 9). The upward stream of hot exhaust gases visible in Figures 7, 9 and 10 should therefore involve the use of a protective layer of refractory material, e.g., chamotte, etc. The geometry of the proposed combustion chamber thus ensures a long residence time (from 1 to 2 s) of gas particles in the high temperature zone (Figure 10). Figure 11 shows numerical analysis of deflector applications that is often used by boiler manufacturers. The obtained results clearly indicate that, in this way, the effect of flue gas swirling and its internal circulation is reduced, which is favourable for the low-emission nature of the boiler operation. Therefore, in the structure proposed by the authors, this classic solution should be abandoned. The risk of high temperatures affecting the top wall of the chamber can be reduced by a protective layer of refractory material.

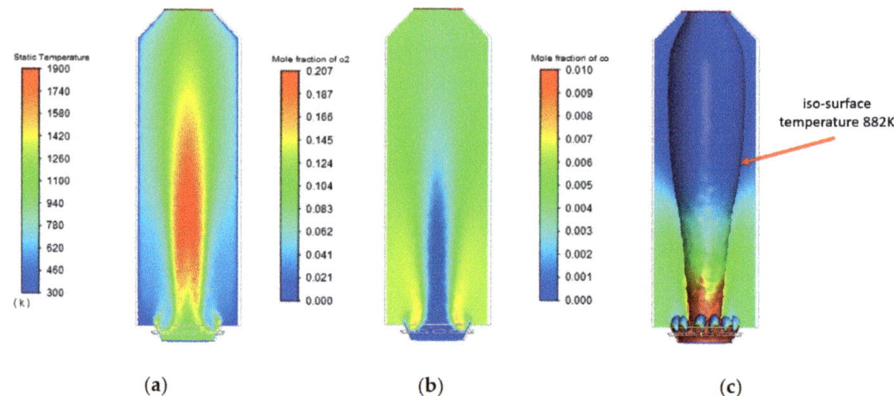

Figure 5. (**a**) Temperature distribution (K); (**b**) O_2 (mole fraction); (**c**) CO (mole fraction) for the 1st phase of numeric modelling.

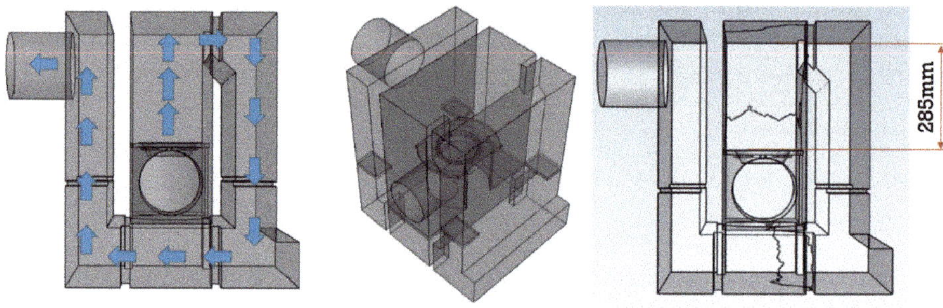

Figure 6. Geometry of furnace modelling.

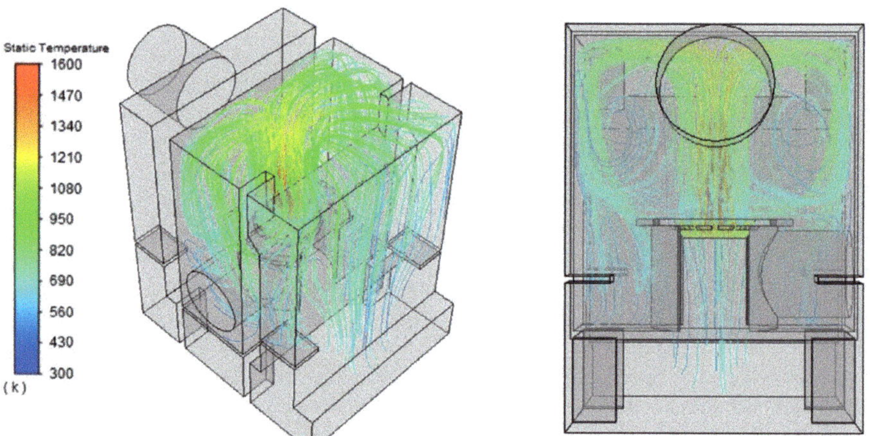

Figure 7. Gas particle paths—temperature scale (K).

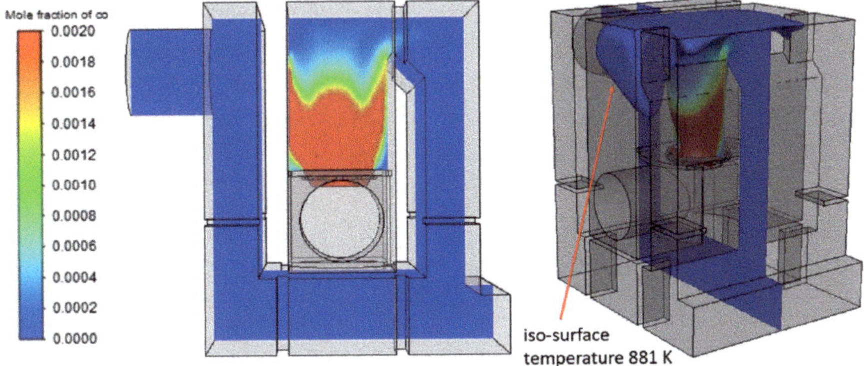

Figure 8. CO concentration (mole fraction).

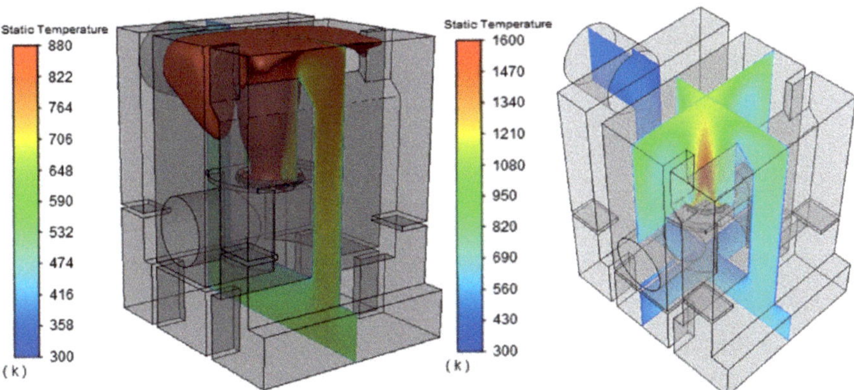

Figure 9. Temperature distribution (K) with determined temperature zone of above 880 K.

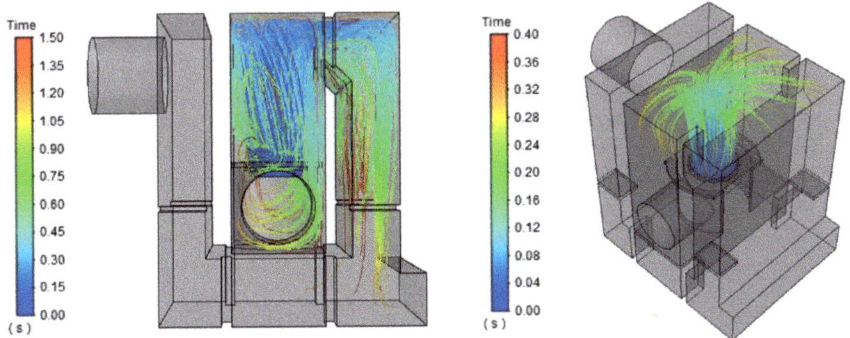

Figure 10. Tracks of gas particles escaping from the burner—time scale (s).

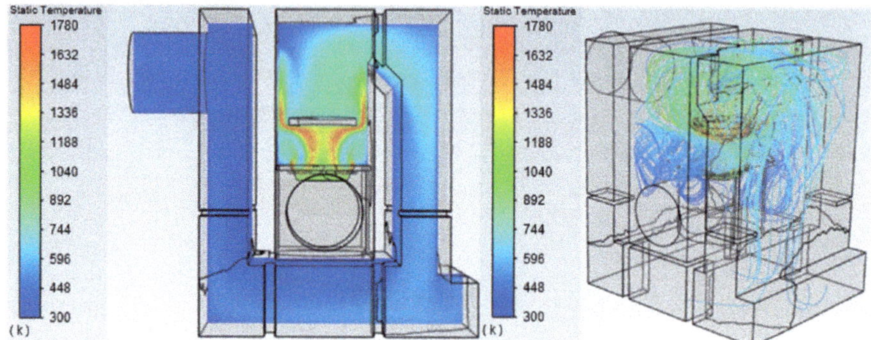

Figure 11. Temperature distribution (K) and gas paths in the boiler model with deflector.

Table 4. Conditions in the chimney flue.

Parameter	Unit	Value
Temperature	K	467
CH_4	% of mole fraction	0.0
CO	% of mole fraction	0.01
O_2	% of mole fraction	10.5
H_2O	% of mole fraction	1.6
CO_2	% of mole fraction	10.9

3.2. Real Model of the Boiler

Based on the results obtained from the numerical model of syngas combustion in the retort burner and the numerical model of the boiler, a prototype of a 10 kW boiler with a retort burner, a centrally located combustion chamber, and exhaust gas circulation around the combustion chamber were constructed. The real model of the boiler is shown in Figure 12. The boiler is built of the combustion chamber, where the burner (6) burning fuel in the form of wood pellets is placed at the bottom (the fuel characteristics are presented in Table 1). The furnace is fenced from below from the flue gas circulation chamber (4) and the convection chamber I (2) and the convection chamber II (3) slot parts. It is fenced off by a sliding cast iron or steel ceramic grate with heat-resistant properties (5) located under the burner (6). The grate (5) tightly closes the furnace (1) so that the fumes are not drawn downwards to the circulation chamber (4). In order to remove the ash from biomass gasification, the grate (5) is manually extended, and after removing the ash, it is moved to a tight position. The ash is poured into chamber (4), from where it is removed periodically outwards through the cleaning duct (8). Hot flue gases cool due to natural draft. They then flow from the furnace, through the convection, the flue gas circulation chamber (4) and the convection chamber II, into the smoke conduit (11) and escape to the atmosphere through the chimney. The chimney draft is regulated by a damper (7). The combustion and convection chambers are enclosed with a water jacket (9) (flue gas/water heat exchanger). The exhaust gas circulation process (which is significant) is forced by the temperature difference in the natural course of the furnace (1), convection chamber I (2), recirculation chamber (4) (exhaust gas recurrence) under the burner (6), and convection chamber II (3). Cooled flue gas is directed to the chimney via the smoke conduit (11). Fuel to the retort burner is dispensed from the tank through stub (10). Inspection hatches (12) and (13) periodically allow observation and removal of impurities that can be deposited on the heating surfaces of the furnace (1) and convection chambers (2) and (3). It should be added that, during stable boiler operation, no intrusive sediment is present on these surfaces. In addition to the exhaust gas circulation process, the central location of the combustion chamber (1) is noteworthy.

This design solution enables the combustion process in chamber (1) to be carried out in adiabatic conditions with a uniformly high temperature distribution, which promotes the efficient combustion of flammable gaseous particles and soot. An expression of thermal processes occurring in the combustion chamber is the absence of tarry (black) deposits on the boiler's heating surfaces. The flue gas/water heat exchange surface is ca. 1.2 m^2, which translates into relatively small external dimensions of the boiler 10 kW − 700 × 780 × 550 mm (Figure 12).

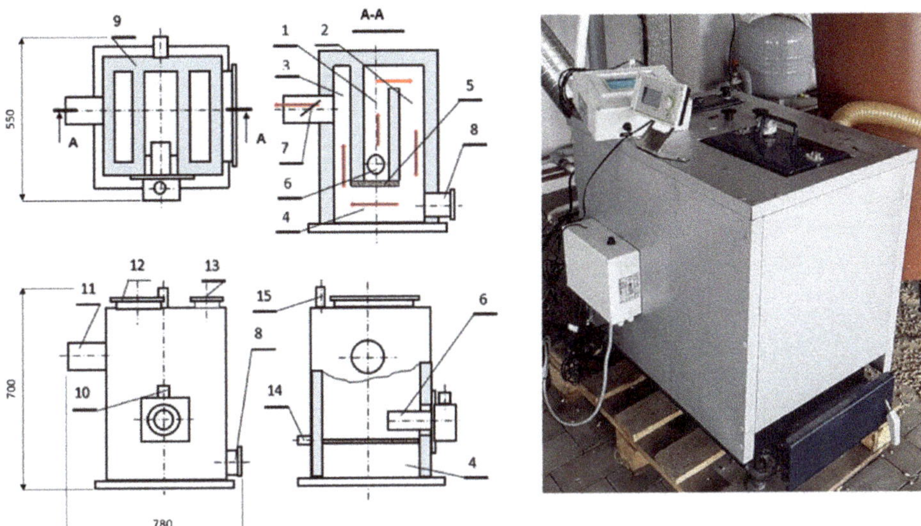

Figure 12. View of the boiler with 10 kW flue gas circulation: 1—combustion chamber, 2—convection chamber I, 3—convection chamber II, 4—combustion gas circulation chamber, 5—airtight sliding grate closing the combustion chamber, 6–10 kW retort burner, 7—thrust adjustment slide, 8—ash removal cleaning duct, 9—water jacket (flue gas/water heat exchanger), 10—fuel feed (pellet), 11—flue (exhaust outlet), 12—inspection hatch of the furnace and convection chamber II, 13—inspection hatch for convection chamber II, 14—return water connector, and 15—feed water connector.

The fuel stream was measured several times using a weighing device. Temperature in the combustion chamber was measured about 0.10 m above the flame with a radiation shielded thermocouple PtRhPt connected additionally with a temperature meter for value comparison. Heat received by the boiler water and boiler heat output were measured with an ultrasonic heat meter. All the obtained data (measured continuously) were transmitted to a personal computer via a data acquisition system. For each test run, parameter values were collected every 5 s, and averaged and mean values were calculated. This time interval of data gathering was optimal, since the measured values were not changing very quickly. Mean values for several consecutive test runs were used to obtain the overall mean value for each type of experiment. Uncertainty intervals were calculated for all measurement results with a 0.95 probability. Pellets were supplied from the hopper by means of a fixed-speed screw feeder and gravitationally fell into a small chamber. Subsequently, a horizontal fixed-speed screw pellet dispenser of each furnace introduced pellets into the burning region. Both devices were synchronized and always worked simultaneously. The fuel stream could be regulated manually by modifying operation and stand-by time of the screw pellet dispenser. The pellet furnace was equipped with its own electrically heated automatic ignition device. Air supply was provided to the furnace by a fan integrated with the furnace. The boiler lacked an automatic device with an oxygen probe (lambda sensor) that would measure oxygen concentration in the flue gas downstream the boiler. Air stream could be modified manually by fan speed regulation, ranging from

10 to 100% of its maximum value. Secondary air entered the furnace through a perforated plate located at the bottom of pellet furnace. The fan stopped operating when boiler water temperature reached its maximum value of 85 °C and reinitiated after the temperature decreased 5 °C below this maximum value.

3.3. Verification of the Numerical Model in Real Conditions

In experiments carried out with the boiler, we strived to achieve a steady state in order to analyse the behaviour of the boiler. Parameters such as temperature and CO_2, O_2 and CO emissions were compared with those collected from numerical simulations, performed using the model (Figures 13 and 14). Analysing the data of the average temperature values in the chimney and CO_2, O_2 emissions from a 2-h measurement and analogous values obtained from the numerical simulation, one can notice differences in the range of 2.7–7%. Experimental data show slight oscillations of the measured quantities, which is a derivative of the nature of a pellet-operated heating installation. Steady-state results are obtained from the numerical model. This is the reason for slight differences in values, but it does not affect the general nature of the simulation results obtained, based on which the research was continued. The boiler geometry adopted for numerical calculations was mapped (Figure 6), the design (executive) documentation was prepared, and the prototype presented in Figure 12 was constructed. The fuel used was wood pellets with the fuel characteristics given in Table 1. The results of the wood pellet combustion tests refer to the working boiler in optimal, determined, full load conditions, i.e., = 10 kW. Under these conditions, the boiler obtained the highest efficiency. It was, on average, η = 92.6% at the fuel calorific value LHV = 16,165 kJ/kg, with fuel consumption 0.000668 g/s and the excess air ratio was λ = 2. The efficiency estimation was based on Formula (7)

$$\eta = \frac{\dot{Q}}{\dot{B} \cdot LHV} \tag{7}$$

where $\dot{Q}$—heat stream generated in the boiler (kJ/s), $\dot{B}$—fuel stream (kg/s), *LHV*—calorific value of fuel (kJ/kg), η—boiler energy efficiency.

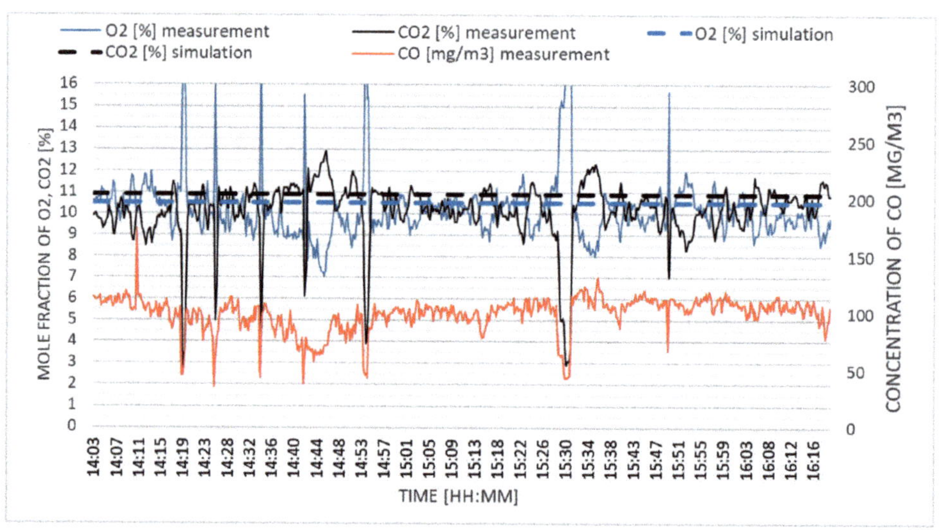

Figure 13. Measured values of gaseous combustion products emission CO, CO_2, and O_2.

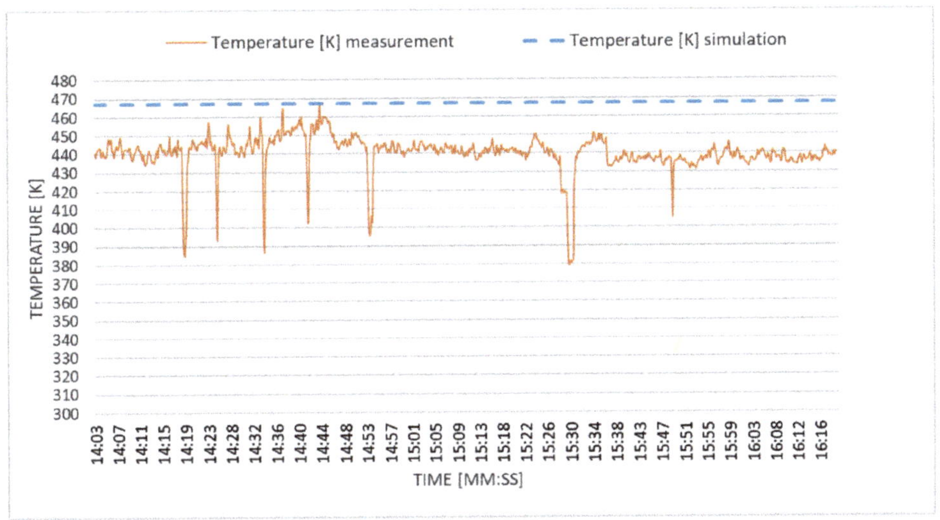

Figure 14. The measured value of the flue gas temperature in the chimney duct.

For example, in six attempts to burn wood pellets (when the percentage of oxygen and carbon dioxide in the waste gas was $O_2 = 10.7\%$ and $CO_2 = 9.93\%$), the content in the carbon monoxide flue gas was on average $CO = 91$ mg/m^3. The emission values obtained through the experiment confirmed the reliability of the previously obtained results using the numerical method presented in Figures 5 and 8. The obtained emission and efficiency parameters result from the previously carried out numerical analysis of combustion in the retort burner shown in Figure 2.

The prototype of the boiler proposed in this paper is characterized (according to numerical calculations) by a simple structure with a wide possibility of modernization in the direction of the following: (i) obtaining higher energy efficiency; (ii) low NO_x emissions; (iii) increasing power, e.g., to 30 kW by increasing the flue gas/water heat exchange surface with unchanged dimensions; and, (iv) increasing the efficiency of the boiler in condensing operation with the possibility of adding (in a simple way) an additional heat exchanger condensing water in the flue gas.

The temperature measurement in the chimney duct (Figure 14) indicated a value of 430 K to 470 K (157–197 °C) and the temperature measurement in the combustion chamber indicated 723 K to 773 K (450–500 °C). Both values were confirmed by numerical calculations.

Figures 13 and 15 show the concentrations of the gaseous components of the flue gas CO, CO_2, O_2, and NO_x measured in the flue (Figure 4) during the 2 h of boiler operation. The low emission of the boiler (which is related to the achievement of complete combustion conditions in it) refers to low concentrations of CO emissions, the values of which are approximately 100 mg/Nm^3. Low CO emission is associated with the proposed combustion chamber system and the circular circulation of the exhaust gases in the convection channels. Such a thermal-flow system ensures swirling of exhaust gases and turbulence, which promotes the mixing of combustible compounds and oxygen under high temperature conditions. Under these conditions, the instantaneous NO_x concentration values were in the range of 100–300 mg/Nm^3, while the average value was 197 mg/Nm^3. It was consistent with the simulation results, i.e., the value of 187 mg/Nm^3. Low, similar values of NOx concentration can be obtained by the classic air and fuel staging method [22,23]. Figure 16 shows a simulation of the NO_x emission depending on the excess air factor (λ) and the nominal boiler power. It should be noted that the proposed combustion technique allows one to obtain average values of NO_x concentrations below 200 mg/Nm^3 (converted into 10% of oxygen in the flue gas) for the nominal power of 10 kW and to ensure parameters are in line with the requirements of the ECODESIGN Directive. This is confirmed, not only by the simulation results, but also by experimental measurements. Even with a power cut of 40%, NO_x emission remains stable. Increasing the excess air factor to 2.5 should enable a further reduction of NO_x emissions by lowering the flame temperature and limiting the thermal mechanism of nitrogen oxide formation. It should be emphasized that, with restrictive exhaust emission standards, it seems fully justified to equip boiler installations with a responsive mechanism of reacting to instantaneous emissions by adjusting operating parameters to temporary operating conditions.

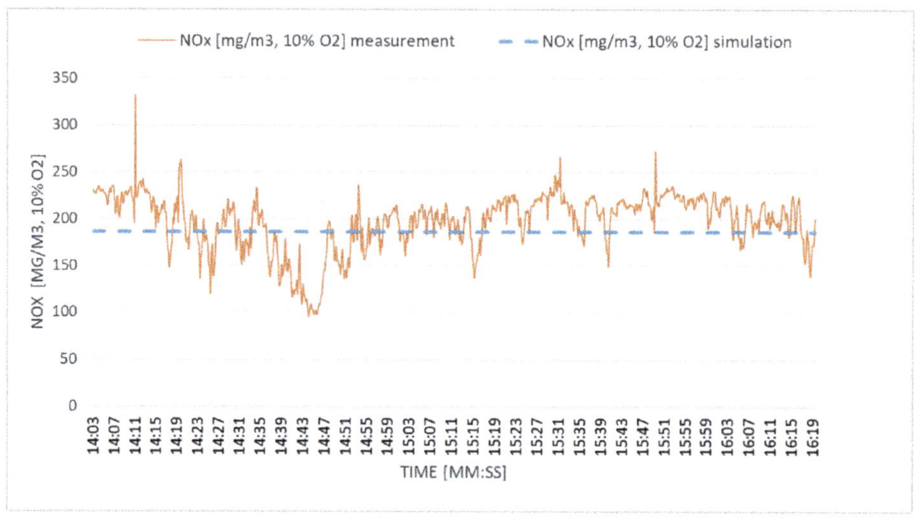

Figure 15. Emissions of nitrogen oxides NO_x measured in the chimney.

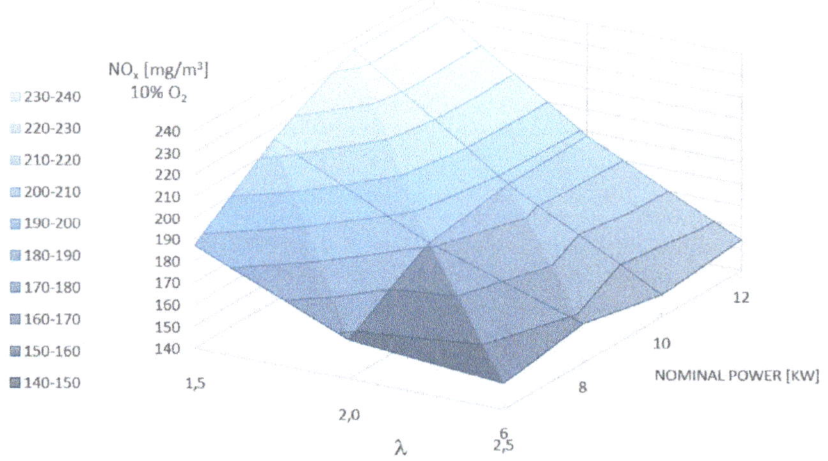

Figure 16. Numerical simulation of the combined effect of nominal power and λ on NO_x concentration in the flue gases.

4. Conclusions

The presented research works aimed at building a prototype of a high-efficiency and low-emission heating boiler with a capacity of 10 kW included the following:

(1)   Experimental and numerical modelling of the wood pellet combustion process in a retort burner,

(2)   Numerical modelling of thermal and flow problems in the furnace and convection flue channels, and

(3)   Experimental verification of the results obtained in the numerical model.

As a result of the performed work, it was proved that the values according to the adopted numerical models turned out to be consistent with the results obtained on the test stand (where the efficiency and emission tests of the boiler prototype were carried out). The obtained results gave credibility to the innovative concept of boiler construction adopted in the numerical model. A characteristic feature of the innovative design is the centrally located irradiated combustion chamber (furnace) and the circulation of hot exhaust gases in convection channels arranged around the chamber. Such arrangement of the furnace and convection channels ensures (which has been demonstrated numerically and experimentally) quasi-adiabatic combustion at a constant temperature. In the entire volume of the combustion chamber (especially in its upper part and the first convection sequence), it ensures effective mixing of hot exhaust gases with air. This results in low concentrations in the exhaust gases (CO—91 mg/Nm^3 on average, NO$_x$—197 mg/Nm^3) and is conducive to achieving high efficiency, the average value of which is 92%. The given values refer to the 10% share of O$_2$ in the flue gas. Noteworthy is the low, allowable (according to ECODESIGN) NO$_x$ emission value. With a power cut of 40%, NO$_x$ emission remains stable. Increasing the excess air factor to 2.5 should enable a further reduction of NO$_x$ emissions by lowering the flame temperature and limiting the thermal mechanism of nitrogen oxide formation. It should be emphasized that, with restrictive emission standards, it seems fully justified to equip boiler installations with a responsive mechanism of reacting to instantaneous emissions by adjusting operating parameters to temporary operating conditions. The obtained results showed that the adopted innovative, non-standard boiler structure with a capacity of 10 kW can be successfully applied for low-emission heating of residential buildings.

The results obtained from the tests of the 10 kW boiler (prototype) indicate the possibility of further research on improving the structure, e.g.,: (i) maintaining its dimensions by increasing the heating surface in convection channels, e.g., with ribbing, placing a bundle of water pipes in the space above the combustion chamber; and, (ii) the reduction of NOx emissions by returning portion of the hot exhaust gases from e.g., the lower (horizontal) convection duct to the combustion chamber.

Author Contributions: Conceptualization, P.M., D.K. and S.P.; Formal analysis, D.K. and S.P.; Investigation, M.J.; Methodology, P.M., D.K. and S.P.; Software, P.M. All authors have read and agreed to the published version of the manuscript.

Funding: This research received no external funding.

Conflicts of Interest: The authors declare no conflict of interest.

References

1. Olsen, Y.; Nøjgaard, J.K.; Olesen, H.R.; Brandt, J.; Sigsgaard, T.; Pryor, S.C.; Ancelet, T.; Viana, M.d.M.; Querol, X.; Hertel, O. Emissions and source allocation of carbonaceous air pollutants from wood stoves in developed countries: A review. *Atmos. Pollut. Res.* **2020**, *11*, 234–251. [CrossRef]
2. Ding, S.; Dang, Y.G.; Li, X.M.; Wang, J.J.; Zhao, K. Forecasting Chinese CO2 emissions from fuel combustion using a novel grey multivariable model. *J. Clean. Prod.* **2017**, *162*, 1527–1538. [CrossRef]
3. Yatkin, S.; Gerboles, M.; Belis, C.A.; Karagulian, F.; Lagler, F.; Barbiere, M.; Borowiak, A. Representativeness of an air quality monitoring station for PM2.5 and source apportionment over a small urban domain. *Atmos. Pollut. Res.* **2020**, *11*, 225–233. [CrossRef] [PubMed]
4. Letschert, V.; Desroches, L.B.; Ke, J.; McNeil, M. Energy efficiency—How far can we raise the bar? Revealing the potential of best available technologies. *Energy* **2013**, *59*, 72–82. [CrossRef]
5. Polverini, D. Energy efficient ventilation units: The role of the Ecodesign and Energy Labelling regulations. *Energy Build.* **2018**, *175*, 141–147. [CrossRef]
6. EUR-Lex—32009L0125—EN—EUR-Lex. Available online: https://eur-lex.europa.eu/legal-content/EN/ALL/?uri=CELEX%3A32009L0125 (accessed on 8 October 2020).

7. Dafnomilis, I.; Hoefnagels, R.; Pratama, Y.W.; Schott, D.L.; Lodewijks, G.; Junginger, M. Review of solid and liquid biofuel demand and supply in Northwest Europe towards 2030 – A comparison of national and regional projections. *Renew. Sustain. Energy Rev.* **2017**, *78*, 31–45. [CrossRef]

8. Nakomcic-Smaragdakis, B.; Cepic, Z.; Dragutinovic, N. Analysis of solid biomass energy potential in Autonomous Province of Vojvodina. *Renew. Sustain. Energy Rev.* **2016**, *57*, 186–191. [CrossRef]

9. Malico, I.; Nepomuceno Pereira, R.; Gonçalves, A.C.; Sousa, A.M.O. Current status and future perspectives for energy production from solid biomass in the European industry. *Renew. Sustain. Energy Rev.* **2019**, *112*, 960–977. [CrossRef]

10. Ferreira, S.; Monteiro, E.; Brito, P.; Vilarinho, C. Biomass resources in Portugal: Current status and prospects. *Renew. Sustain. Energy Rev.* **2017**, *78*, 1221–1235. [CrossRef]

11. Thomson, H.; Liddell, C. The suitability of wood pellet heating for domestic households: A review of literature. *Renew. Sustain. Energy Rev.* **2015**, *42*, 1362–1369. [CrossRef]

12. Verma, V.K.; Bram, S.; Delattin, F.; De Ruyck, J. Real life performance of domestic pellet boiler technologies as a function of operational loads: A case study of Belgium. *Appl. Energy* **2013**, *101*, 357–362. [CrossRef]

13. Büchner, D.; Schraube, C.; Carlon, E.; von Sonntag, J.; Schwarz, M.; Verma, V.K.; Ortwein, A. Survey of modern pellet boilers in Austria and Germany—System design and customer satisfaction of residential installations. *Appl. Energy* **2015**, *160*, 390–403. [CrossRef]

14. Patiño, D.; Crespo, B.; Porteiro, J.; Míguez, J.L. Experimental analysis of fouling rates in two small-scale domestic boilers. *Appl. Therm. Eng.* **2016**, *100*, 849–860. [CrossRef]

15. Sungur, B.; Topaloglu, B. An experimental investigation of the effect of smoke tube configuration on the performance and emission characteristics of pellet-fuelled boilers. *Renew. Energy* **2019**, *143*, 121–129. [CrossRef]

16. Collazo, J.; Porteiro, J.; Míguez, J.L.; Granada, E.; Gómez, M.A. Numerical simulation of a small-scale biomass boiler. *Energy Convers. Manag.* **2012**, *64*, 87–96. [CrossRef]

17. *User Manual for the Synthesis Gas Analyzer GAS 3100P*; G.E.I.T. EUROPE: Bunsbeek, Belgium, 2019.

18. *ANSYS®Fluent, Release 2019 R2, Help System*; ANSYS, Inc.: Canonsburg, PA, USA, 2019.

19. Chapela, S.; Porteiro, J.; Garabatos, M.; Patiño, D.; Gómez, M.A.; Míguez, J.L. CFD study of fouling phenomena in small-scale biomass boilers: Experimental validation with two different boilers. *Renew. Energy* **2019**, *140*, 552–562. [CrossRef]

20. Gómez, M.A.; Martín, R.; Chapela, S.; Porteiro, J. Steady CFD combustion modeling for biomass boilers: An application to the study of the exhaust gas recirculation performance. *Energy Convers. Manag.* **2019**, *179*, 91–103. [CrossRef]

21. Drosatos, P.; Nesiadis, A.; Nikolopoulos, N.; Margaritis, N.; Grammelis, P.; Kakaras, E. CFD Simulation of Domestic Gasification Boiler. *J. Energy Eng.* **2017**, *143*, 04016052. [CrossRef]

22. Liu, H.; Chaney, J.; Li, J.; Sun, C. Control of NOx emissions of a domestic/small-scale biomass pellet boiler by air staging. *Fuel* **2013**, *103*, 792–798. [CrossRef]

23. Chaney, J.; Liu, H.; Li, J. An overview of CFD modelling of small-scale fixed-bed biomass pellet boilers with preliminary results from a simplified approach. *Energy Convers. Manag.* **2012**, *63*, 149–156. [CrossRef]

Publisher's Note: MDPI stays neutral with regard to jurisdictional claims in published maps and institutional affiliations.

Article

Low-Cost Organic Adsorbents for Elemental Mercury Removal from Lignite Flue Gas

Marta Marczak-Grzesik [1,2,*], Stanisław Budzyń [1], Barbara Tora [3], Szymon Szufa [4] , Krzysztof Kogut [1] and Piotr Burmistrz [1]

1 Faculty of Energy and Fuels, AGH University of Science and Technology, Mickiewicz Avenue 30, 30-059 Krakow, Poland; budzyn@agh.edu.pl (S.B.); kogut@agh.edu.pl (K.K.); burmistr@agh.edu.pl (P.B.)
2 AGH Centre of Energy, AGH University of Science and Technology, Czarnowiejska 36, 30-054 Krakow, Poland
3 Faculty of Mining and Geoengineering, AGH University of Science and Technology, Mickiewicz Avenue 30, 30-059 Krakow, Poland; tora@agh.edu.pl
4 Faculty of Process and Environmental Engineering, Lodz University of Technology, Wolczanska 213, 90-924 Lodz, Poland; szymon.szufa@p.lodz.pl
* Correspondence: mmarczak@agh.edu.pl

Abstract: The research presented by the authors in this paper focused on understanding the behavior of mercury during coal combustion and flue gas purification operations. The goal was to determine the flue gas temperature on the mercury emissions limits for the combustion of lignites in the energy sector. The authors examined the process of sorption of mercury from flue gases using fine-grained organic materials. The main objectives of this study were to recommend a low-cost organic adsorbent such as coke dust (CD), corn straw char (CS-400), brominated corn straw char (CS-400-Br), rubber char (RC-600) or granulated rubber char (GRC-600) to efficiently substitute expensive dust-sized activated carbon. The study covered combustion of lignite from a Polish field. The experiment was conducted at temperatures reflecting conditions inside a flue gas purification installation. One of the tested sorbents—tire-derived rubber char that was obtained by pyrolysis—exhibited good potential for Hg^0 into Hg^{2+} oxidation, resulting in enhanced mercury removal from the flue. The char characterization increased elevated bromine content (mercury oxidizing agent) in comparison to the other selected adsorbents. This paper presents the results of laboratory tests of mercury sorption from the flue gases at temperatures of 95, 125, 155 and 185 °C. The average mercury content in Polish lignite was 465 $\mu g \cdot kg^{-1}$. The concentration of mercury in flue gases emitted into the atmosphere was 17.8 $\mu g \cdot m^{-3}$. The study analyzed five low-cost sorbents with the average achieved efficiency of mercury removal from 18.3% to 96.1% for lignite combustion depending on the flue gas temperature.

Keywords: lignite; anthropogenic emission; mercury removal; flue gases purification; low-cost asorbents

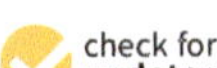

check for **updates**

Citation: Marczak-Grzesik, M.; Budzyń, S.; Tora, B.; Szufa, S.; Kogut, K.; Burmistrz, P. Low-Cost Organic Adsorbents for Elemental Mercury Removal from Lignite Flue Gas. *Energies* **2021**, *14*, 2174. https://doi.org/10.3390/en14082174

Academic Editor: S.M. Ashrafur Rahman

Received: 15 March 2021
Accepted: 11 April 2021
Published: 13 April 2021

Publisher's Note: MDPI stays neutral with regard to jurisdictional claims in published maps and institutional affiliations.

1. Introduction

Ecotoxic elements, especially mercury, are particularly hazardous substances among pollutants with no physiological relevance for living organisms. It is generally known that mercury exposure can inflict various health issues, especially neurological, immunological, behavioral and sensory issues [1,2]. The mentioned afflictions were diagnosed in consumers of contaminated fish (Minamata Disease in Japan) and crops (Iraq, Guatemala and Russia). Mercury has an adverse impact on health, as well as long atmospheric lifetime and propensity for deposition in the aquatic environment and in living tissue. Due to the global distribution of mercury, the US Environmental Protection Agency [3] has classified it and its compounds as an air quality threat. Coal combustion release, which has constantly increased following growing worldwide energy demand, is considered as one of the most significant anthropogenic origins of mercury [3–5]. Therefore, due to the element's (wide) exposition, to avoid adverse harmful effects to the respiratory, nervous and immune systems, it is crucial to decrease atmospheric mercury emissions. Conducted research confirms

49

the severe influence on the environment and warrants worldwide actions for the reduction of emissions [6–8].

In Poland, solid fuel combustion, mainly coal, constitutes the dominating source of mercury emission, exceeding 80% of the country's share [9,10]. Under the adoption of 2010/75/UE Directive (IED—Industrial Emissions Directive) that includes the permitted industrial emissions, the Polish energy sector has introduced the Best Available Technologies (BAT) methods with a permissible emission limit of 1–10 $\mu g \cdot Nm^{-3}$, depending on the fuel used and plant size [10,11]. At present, the vast majority of units in Polish power plants do not meet the BAT-defined standards for mercury, which imposes the need for technological progress. As new emerging technologies such as clean coal technologies or deep desulfurization and denitrogenation do not meet the European standards for mercury emission, auxiliary means of mercury removal are needed to prevent power plant closures due to non-compliance with regulations. Among the techniques of standalone mercury removal systems, injection of powdered activated carbon (PAC) to flue gasses due to the high specific surface of the material is commonly utilized in the United States of America [12,13].

The efficiency of mercury removal using powdered sorbent injection depends on both physical and chemical properties of the material, flue gas temperature (inversely proportional, i.e. increases in temperature result in sorption decreases) and flue gas constituents (presence of halogen compounds and sulfur trioxide (SO_3)). Furthermore, mercury speciation plays a vital role as the Hg^{2+} form of mercury has good affinity for sorbents capture, whereas Hg^0 is practically not adsorbed. Therefore, oxidation of the mercury in order to increase the share of Hg^{2+} is an established practice [14–17].

The main disadvantage of activated carbon sorbent usage is the material price, which results in global research [18–21] aimed at the search for new low-cost adsorbents with comparable sorptive properties that can be used as a replacement (low-cost adsorbents). The authors suggest that the use of waste materials with no industrial application can be the most beneficial in both economic and environmental terms [22]. For example, Guangqian L. et al. [23] developed waste-derived sorbents from biomass and brominated flame retarded plastic for mercury removal from coal-fired flue gas. Charpenteau C. et al. [24] proposed coal fly ash as low-cost material. The authors of [25,26] analyzed six low-cost sorbents with the average achieved efficiency of mercury removal of 30.6–92.9% for sub-bituminous coal and 22.8–80.3% for lignite combustion. One of the main objectives of this study was to recommend a low-cost organic sorbent such as coke dust to efficiently substitute expensive activated carbon for mercury removal from flue gas. Other researchers have attempted to develop alternative low-cost yet efficient adsorbents utilizing agricultural and industrial wastes [27–29]. Biochars play a significant role in addressing the current demands of adsorbents for various applications [30,31]. Initial research on specially prepared chars from rubber wastes has proven beneficial for the capture of both mercury [32] and other heavy metals [33]. Considering the amount of mercury released by the energy sector [34–40], the research was organized to determine the potential of rubber char as a sorbent for capturing mercury from flue gas at different temperatures and compare it to other possible adsorbents as an alternative for expensive activated carbons.

The purpose of this study was the application of dust-sized sorbents for reduction of mercury emissions from flue gas. The main objective of this study was to recommend a low-cost organic adsorbent such as coke dust (CD), corn straw char (CS-400), brominated corn straw char (CS-400-Br), rubber char (RC-600) or granulated rubber char (GRC-600) to efficiently substitute expensive dust-sized activated carbon. The study covered combustion of lignite from a Polish field. The experiment was conducted at temperatures reflecting conditions inside a flue gas purification installation.

2. Materials and Methods

2.1. Sorbents Origins

To compare the efficiency of sorbents, studies consisted of five types of low-cost organic sorbents in three different forms: coke-derived char dust (CD), biomass chars (CS, CS-Br) and rubber-derived chars (RC and GRC). Commercially available activated carbon (AC) was included as industrial flue gas mercury sorbent of choice for comparative reasons. The sorbents used are described in Table 1.

Table 1. Description of low-cost organic adsorbent.

LCOA	Sign	Description of Origin
Activated carbon	AC	Commercial activated carbon, dedicated, e.g., for gas-phase mercury removal. It was formed in the process of coal carbonization and subsequent thermal activation of the obtained structure.
Coke dust	CD	Byproduct of large-scale coke production. The dust is obtained during coke dry-cooling process, hauling and sorting.
Corn straw char	CS-400	Solid product (char) of corn straw torrefaction process using superheated steam at 400 °C.
Brominated Corn straw char	CS-400-Br	Solid product (char) of corn straw torrefaction process with the addition of bromine; blend char/Br2 was prepared at ratio 5:1.
Rubber Char	RC-600	Solid product (char) of car tire pyrolysis at 600 °C. The material was derived from industrial installations in which intact tires are subjected to high temperatures.

LCOA, Low-Cost Organic Adsorbent.

2.2. Sorbents Analysis

The scope of the sorbent analysis listed in Section 2.1 includes:

(i) Proximate and ultimate analyses were performed in accordance with the ISO standard [41,42].

(ii) Determination of chlorine content was evaluated as chlorine anion content in water solution using a direct reading spectrophotometer (DR/2000 HACH). A sample was combusted in AC-350 bomb calorimeter (LECO) with Eschka mixture—in accordance with the ISO standard [43].

(iii) Mercury content was analyzed by thermal decomposition, amalgamation and atomic absorption spectrophotometry (DMA-80 Direct mercury analyser).

(iv) The particle size of LCOA was analyzed by ISO standard [44].

(v) The porous texture of all samples was analyzed using nitrogen adsorption–desorption at 77 K using Autosorb®-1-C (Quantachrome Instruments, USA) according to the standards [45–48].

(vi) Bromine content was analyzed with X-ray spectrometry with wavelength dispersion in PROMUS II sequential spectrometer (Rigaku).

(vii) ED-XRF analyses of samples in a powder form were conducted with use of PANalytical EmpyreanXLE diffractometer with copper anode (Cu Kα) in the 2θ angle range of 10–110°. The type and amount of crystalline phase were evaluated with PANalytical HighScore Plus software.

Values of these parameters were determined for the air-dried basis of the sample.

2.3. Mercury Adsorption System

The stand for measuring mercury sorption from flue gases generated by the combustion of solid fuels is shown in Figure 1. The bench-scale measurement setup consists of tube furnace (3) along with temperature and gas flow regulation with quartz tube (4) as the combustion chamber, gas source (1) and sorbent holder (8). The utilized equipment allows for controllable gas heating at specified points (9). The measurement procedure involves air-fueled coal sample (5) combustion and analysis of flue-gas mercury capture by the sorbent.

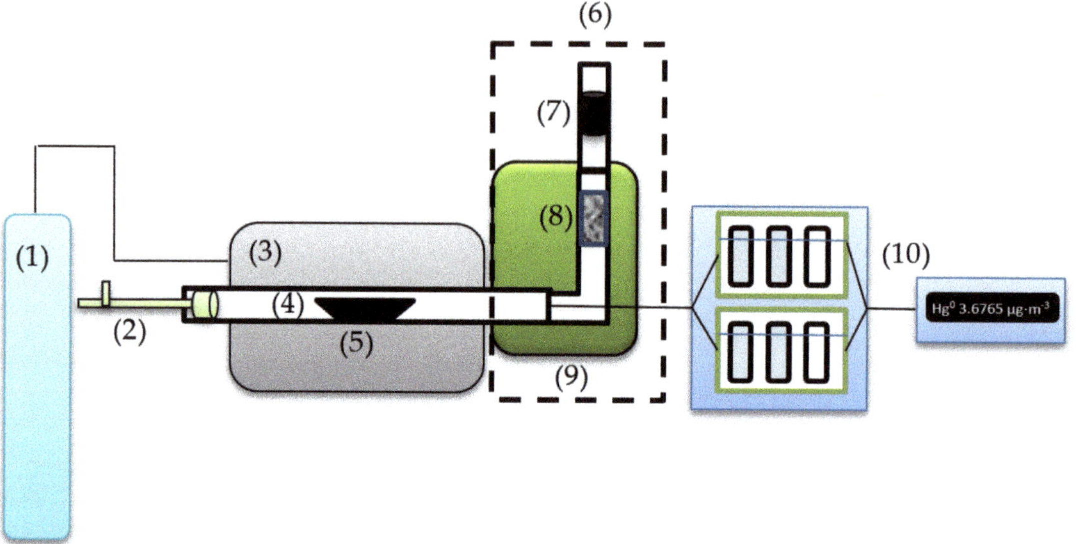

Figure 1. Test stand schematic: (**1**) gas source, air; (**2**) metal rod; (**3**) tube furnace; (**4**) quartz tube; (**5**) coal sample; (**6**) flue mercury adsorption device; (**7**) sorbent trap; (**8**) sorbent holder; (**9**) flue gases heating furnace; and (**10**) set of impingers with mercury survey meter EMP-3 of impingers for mercury speciation measurement.

The experiment was conducted with predetermined temperature conditions for sample combustion and sorbent temperature with similar combustion times and airflow. One gram of the sample was positioned in a ceramic boat-shaped crucible and progressively transported to the center of the combustion zone with a metal rod (2). The standardized measurement time after the crucible was introduced to the center of the combustion zone was 20 min. Additionally, LCOA was placed a sorbent trap (7), which measured the concentration of Hg emitted into the atmosphere.

A more detailed description of the experimental conditions is presented in [25]. For the combustion process, lignite was used. It underwent ultimate and proximate analysis with the additional steps of mercury and chlorine determination in accordance with the methodology described in Section 2.2.

The lignite sample was prepared in accordance with the ISO standards [49]. The characteristic of the used lignite is shown in Section 3.2.

Mercury Speciation Testing

To determine mercury speciation in flue gas samples, two sets of impingers (three impingers in each set) and a mercury survey meter EMP-3 (continuous mercury analyzer) were used. For the period of mercury speciation ascertainment, the flue mercury adsorption device (Figure 1, Point 6) was disengaged. Flue gases flowed by a series of impingers (Figure 1, Point 10). Raw flue gas was channeled into two flow streams. In the first vessel out of the first gas washing bottle setup, the Hg^{2+} mercury from the first stream was reduced to Hg^0 with 10% $SnCl_2$ solution, followed by acidic gas scrubbing with 10% KOH solution and subsequent moisture removal with the third vessel. The purified flue gas stream was then analyzed with an EMP-3 detector for total mercury content determination. The second gas washing bottle setup had an analogous configuration to the first setup, with the difference being the first vessel was used to capture Hg^{2+} mercury out of the second stream with 10% KCl solution. Consequently, the EMP-3 detector analyzed the purified flue gas from the second stream for elementary mercury. The concentration of Hg^{2+} was determined based on the difference between total and elementary mercury.

2.4. Methodology of Mercury Adsorption

To check the reliability of the tests performed, the balance of mercury in the laboratory installation was calculated for each experiment in accordance with the following model:

$$m_c \cdot C_0 - m_{ash} \cdot C_{ash} = m_{sorb} \cdot \left(C''_{Hg} - C'_{Hg} \right) + m_{trap} \cdot \left(C''_{trap} - C'_{trap} \right) \tag{1}$$

where m_c is the mass of combusted coal (kg); C_0 is the mercury content in coal ($\mu g \cdot kg^{-1}$); m_{ash} is the mass of ash from coal combustion (kg); C_{ash} is the Hg content in the ash (kg); m_{sorb} is the weight of tested sorbent (kg); C'_{Hg} and C''_{Hg} are the mercury concentrations in sorbent before and after sorption ($\mu g \cdot kg^{-1}$); m_{trap} is the mass of sorbent trap [kg]; and C'_{trap} and C''_{trap} are the mercury concentrations in the trap before and after the experiment ($\mu g \cdot kg^{-1}$).

Based on the obtained data and measurements of flue gas mercury concentration before and after adsorption, the Hg adsorption capacities (q, $\mu g \cdot g^{-1}$) of the sorbent samples were calculated using below equation:

$$q = \frac{Q_{fg}}{m_{sorb}} \int_0^t (C''_{fg} - C'_{fg}) \, dt \tag{2}$$

where Q_{fg} is the gas flow rate ($m^3 \cdot min^{-1}$) and C''_{fg} and C'_{fg} are the inlet and outlet Hg concentrations in flue gases ($\mu g \cdot Nm^{-3}$) at combustion time t (min).

Additionally, mercury removal efficiency of tested sorbents (MR, %) was calculated as:

$$MR = \frac{C''_{Hg} - C'_{Hg}}{C_0 - C_{ash}} \cdot 100\% \tag{3}$$

3. Results and Discussion

3.1. Sorbent Characteristics

AC contained 6.1 $\mu g\ Hg \cdot kg^{-1}$ (Table 2). Relatively small quantities of mercury were also found in CD (8.9 $\mu g \cdot kg^{-1}$) and CS-400 (2.7 $\mu g \cdot kg^{-1}$) CD and AC were obtained in the carbonization process, therefore they contained minimal quantities of volatiles: 3.2 wt% for CD and 15% for AC. RC-600 and GRC-600 had high mercury content, at 158 and 73 $\mu g \cdot kg^{-1}$, respectively, as well as ash (19.8% and 19.6%, respectively). Commercial activated carbon (AC) and rubber waste chars (RC-600 and GRC-600) recorded from 3.5 to 18 times higher sulfur content than other examined sorbents. Sorbents had bromine content in the range of 50–730 ppm.

Table 2. Properties of the analyzed sorbents.

Sorbent	Proximate Analysis (wt%)			Ultimate Analysis (wt%)			(wt%)		($\mu g \cdot kg^{-1}$)
	M_{ad}	V_{ad}	A_{ad}	C_{ad}	H_{ad}	S_{ad}	Br_{ad}	Cl_{ad}	Hg_{ad}
AC	9.2	15.09	26.2	59.5	1.45	2.11	0.017	0.021	6.1
CD	0.4	3.19	9.8	85.0	0.16	0.59	0.014	0.023	8.9
CS-400	2.6	23.7	22.1	61.7	3.20	0.15	0.005	0.037	2.7
CS-400-Br	-	-	-	-	-	-	0.033	-	2.9
RC-600	1.3	3.2	19.8	75.4	0.91	2.55	0.068	0.063	158.1
GRC-600	2.3	3.2	19.6	76.3	0.95	2.66	0.073	0.075	73.0

M_{ad}, moisture in the air-dried basis; V_{ad}, volatile matter in the air-dried basis; A_{ad}, ash in an air-dried basis; Cl_{ad}, chlorine in the air-dried basis; C_{ad}, carbon in the air-dried basis; H_{ad}, hydrogen in the air-dried basis; S_{ad}, sulfur in the air-dried basis; Br_{ad}, bromine in the air-dried basis; Hg_{ad}, mercury in the air-dried basis.

Coke and rubber waste chars are microporous materials with a moderately developed mesoporous and poor microporous structure (Table 3). The specific surfaces (S_{BET}) of CD,

RC-600 and GRC-600 have several dozen m^2·g^{-1}, while in AC this parameter reaches 670.5 m^2·g^{-1}. The specific surface area of corn straw char is 4.8 m^2·g^{-1}. However, the mesoporous structure is an essential parameter for mercury adsorption because, due to the particle size of mercury, mesopores are considered to be dominant areas of its deposition. RC-600 and GRC-600 sorbents have the highest mesoporous surface among the analyzed samples: 0.38 and 0.17 cm^3·g^{-1}, respectively. Therefore, these samples underwent further qualitative and quantitative phase analysis.

Table 3. Parameters of porous structure for LCOA.

Sample	BET Surface Area (m^2·g^{-1})	Micropore Surface Area (cm^3·g^{-1})	Mesopore Volume (cm^3·g^{-1})	Total Pore (cm^3·g^{-1})
AC	670.5	0.307	0.055	-
CD	24.3	0.006	0.009	0.018
CS-400	4.8	0.001	0.010	0.023
CS-400-Br	4.6	0.001	0.008	0.018
RC-600	70.3	0.025	0.380	0.396
GRC-600	74.7	0.023	0.171	0.239

Qualitative and Quantitative Phase Analysis of Rubber Waste towards as a Sorbent

Based on conducted experimental studies presented in a previous publication [25] sorbents of rubber waste were the most promising material for flue gas mercury capture in lignite combustion processes. Therefore, only this material was selected for Scanning Electron Microscopy (SEM) using Energy Dispersive Spectroscopy (EDS system).

The results of the chemical analysis for selected RC-600 sample points (Figure 2) are shown in Table 4. The char contains 78.5–80.8% carbon, most of which comes from carbon black used for tire production. Due to the point-based nature of the measurement this amount does not correspond to the average content of carbon for the representative sample (Table 2). Another probable origin is tire pyrolytic oil that carbonized during rubber processing. The mineral content in the char is assumed to be a result of industry additives present in the tire manufacturing such as fillers, plasticizers, vulcanizing activators and crosslinking additives. The used matter consists of SiO_2, ZnO CaO, Al_2O_3, Na2O and Fe_2O_3 (Table 4).

Figure 2. SEM image of RC-600.

Table 4. Analysis EDS of selected points of RC-600.

Element	Chemical Components (wt%)		
	1	2	3
C	78.5	86.8	80.9
O	2.0	5.1	5.9
Mg	0.1	0.2	0.2
Al	0.2	0.2	0.3
Si	3.2	1.3	1.4
S	4.7	2.2	2.0
Ca	1.2	0.2	4.8
Co	0.6	0.2	0.3
Cu	1.0	0.2	0.2
Zn	8.4	3.2	2.4

Figure 3 shows example SEM images of the microstructure of RC-600 and GRC-600 chars. The picture presents small carbon particles aggregated into larger formations as the main constituent of the samples.

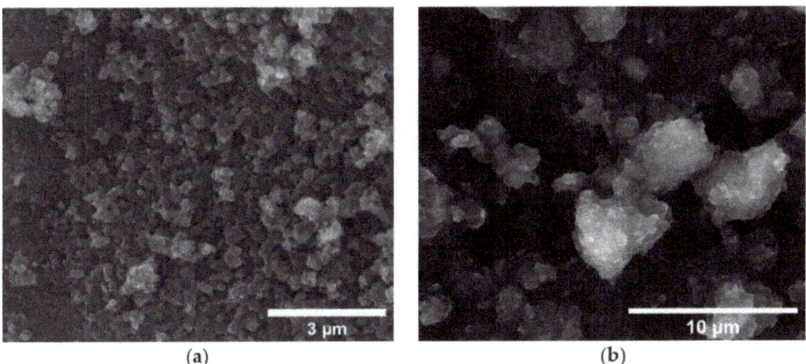

Figure 3. SEM images of: (**a**) RC-600; and (**b**) GRC-600.

3.2. Properties of Lignite

The characteristics of the combusted coal are shown in Table 5. Lignite contained an average of 465 µg·kg^{-1} of mercury. The average content of halogens such as Cl and Br (supporting factors in the oxidation of mercury from Hg^0 to Hg^{2+}) in lignite was equal to 30 ppm Cl and 3.9 ppm Br. The average sulfur content for the lignite was 1.8 wt%; it also contained 23.7 wt% ash, 41.20 wt% volatiles and 12.9 wt% moisture.

Table 5. Characteristic of the coal used in the experiment.

Coal	Proximate Analysis (wt%)			Ultimate Analysis (wt%)			(wt%)	(ppm)		(µg·kg^{-1})
	M_{ad}	V_{ad}	A_{ad}	C_{ad}	H_{ad}	S_{ad}	Ca_{ad}	Br_{ad}	Cl_{ad}	Hg_{ad}
Lignite	12.9	41.20	23.7	43.5	4.90	1.80	2.42	3.9	30	465.0

3.3. Determination Mercury Speciation in Flue Gases during Lignite Combustion

Figure 4 presents a Sankey diagram for mercury released from bench-scale lignite combustion with determination for various species of Hg at the flue gases temperature of 95 °C. Due to the experimental process, ash was a solid residue after coal combustion, and nearly all the mercury contained in the coal changed to the flue gas in gaseous forms (Hg^0 and Hg^{2+}). Only 2% remained in the ashes in the Hgp form. The proportion of Hg^0 in the analyzed flow was relatively high, reaching 70%. The Hg^0:Hg^{2+} ratio (5:2) was determined by the chemical composition of the fuel. Lignite is characterized by low content of chlorine and bromine (Table 5), which leads to oxidation of Hg^0 to Hg^{2+}. This effect, however, can be partially slowed down by relatively high presence of calcium (2.42 wt%), which is capable of chemical deactivation of chlorine due to their chemical affinity [5]. The resulting fuel composition determines the share of Hg^{2+} to be 28% or lower. Due to behavioral differences between both forms, Hg^0 is generally more difficult to remove from flue gas by the adsorption process. Therefore, the determining factor in the selection of sorbent and process conditions for mercury removal should be the chemical composition of the sorbents as well as the sorption temperature.

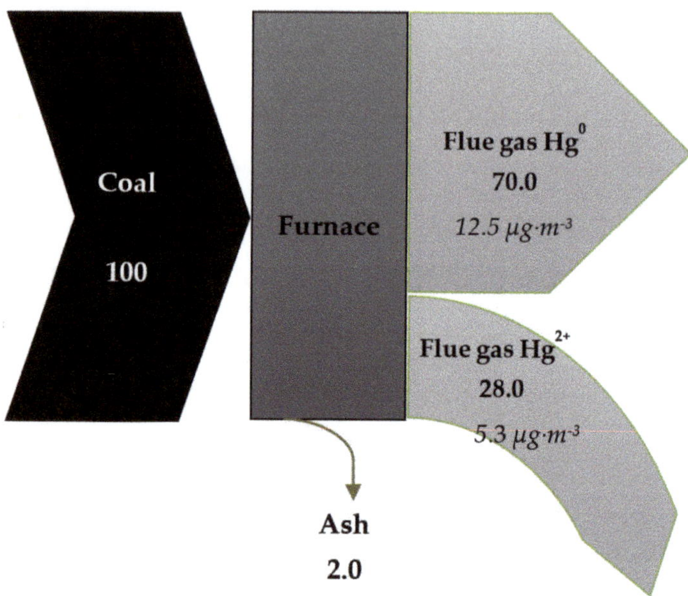

Figure 4. Sankey diagrams of mercury distribution (%).

3.4. Hg Adsorption Performance during Combustion of Lignite

Table 6 shows the Hg adsorption capacity (q) of the tested sorbents and the efficiency of Hg removal from flue gas (MR) at process temperature of 95 °C. The commercial activated carbon, currently used in active flue gas mercury removal methods, was the most efficient. AC removed the mercury almost entirely. The sorption efficiency of CD was also high at 93.8%. RC-600 and GRC-600 presented mercury removal during combustion of lignite of 81.5% and 65.7%, respectively. The analyses showed corn straw char to be the worst sorptive material during lignite combustion. CS-400 decreased the concentration of mercury in flue gas by only 32.4%. CS-400-Br was more efficient in mercury removal (MR at 50%).

Table 6. Efficacy assessment of LCOA for Hg sorption.

Sorbent	q	MR
	$\mu g \cdot g^{-1}$	%
AC	102.6	96.1
CD	100.2	93.8
CS-400	34.3	32.4
CS-400-Br	53.4	50.1
RC-600	87.0	81.5
GRC-600	70.2	65.7

As shown in Figure 5, the Hg concentration in raw flue gas was 17.8 $\mu g \cdot m^{-3}$ (in the process of lignite combustion with a Hg content of 465 $\mu g \cdot kg^{-1}$). Most of the mercury from flue gases was removed by AC and CD: their Hg adsorption capacities were 102.6 and 100.2 $\mu g \cdot g^{-1}$, while the concentration of Hg in flue gases was decreased by 17.1 and 6.7 $\mu g \cdot m^{-3}$, respectively, for AC and CD (Figure 5). Application of corn straw chars resulted in incomplete mercury removal. Hg adsorption capacity for CS-400 and CS-400-Br reached 34.3 and 53.4 $\mu g \cdot g^{-1}$, respectively, which caused the reduction of mercury emission to 12.1 and 8.9 $\mu g \cdot m^{-3}$. The low sorptive capability of CS-400 can be explained by its specific surface of 4.8 $m^2 \cdot g^{-1}$. A higher value of q was obtained for rubber chars; the adsorption capacity was 87.0 $\mu g \cdot g^{-1}$ for RC-600 and 70.2 $\mu g \cdot g^{-1}$ for GRC-600. As a result, Hg concentration in flue gases was reduced to 3.3 and 6.1 $\mu g \cdot m^{-3}$, respectively.

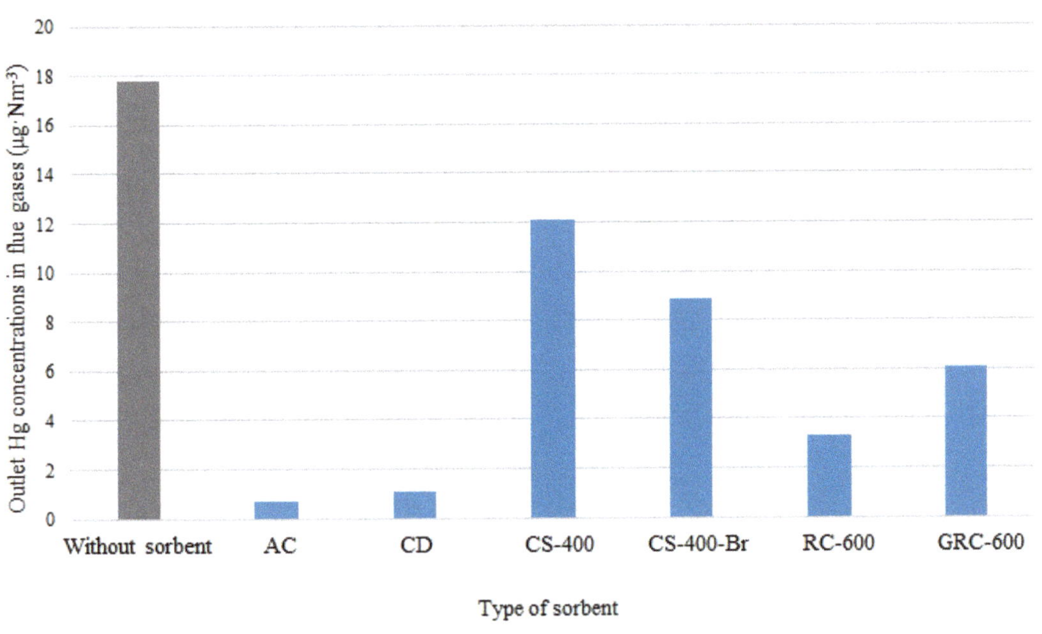

Figure 5. Concentration of mercury in raw flue gas and flue gases cleaned by LCOA.

3.5. Effect of Flue Gas Temperature on Mercury Sorption Ability

The next stage of the study considered the influence of the adsorption temperature on the effectiveness of LCOA in Hg removal. The average mercury removal effectiveness for AC, CD and CS-400 decreased as the temperature increased, with the highest reading at 95 °C and the lowest at 185 °C (Table 7). The change in adsorption temperature had a different effect on RC-600, GRC-600 and CS-400-Br. The average mercury removal efficiency increased as the adsorption temperature increased, with the lowest results at 95 °C and the highest at 185 °C.

Table 7. Efficacy assessment of LCOA for Hg sorption at different temperatures.

Sorbent	Temperature	q	MR
	°C	µg·g^{-1}	%
AC	95	100.2	93.8
AC	125	95.4	89.3
AC	155	73.2	68.5
AC	185	72,0	67.4
CD	95	102,6	96.1
CD	125	94,2	88.2
CD	155	91,8	86.0
CD	185	84,0	78.7
CS-400	95	34.3	32.1
CS-400	125	29.4	27.5
CS-400	155	24.2	22.7
CS-400	185	19.5	18.3
CS-400-Br	95	53.4	50.1
CS-400-Br	125	55.2	51.7
CS-400-Br	155	58.8	55.1
CS-400-Br	185	61.2	57.3
RC-600	95	87,0	81.5
RC-600	125	93,6	87.6
RC-600	155	99,0	92.7
RC-600	185	101,7	95.2
GRC-600	95	70,2	65.7
GRC-600	125	84,0	78.7
GRC-600	155	88,8	83.1
GRC-600	185	97,2	91.0

The average mercury concentration in flue gases during lignite combustion was 17.8 µg·m^{-3} (Figure 6), whereas, after the adsorption process in the temperature range (95–185 °C), the readings showed 1.1–5.8 µg·m^{-3} for AC, 0.7–3.8 µg·m^{-3} for CS and 12.1–14.6 µg·m^{-3} for CS-400. CS-400 presented poor Hg removal performance at all tested temperatures, while the brominated CS-400-Br showed progression in Hg removal efficiency as temperature increased, similarly to RC-600 and GRC-600 (Figure 6).

The deciding factor for Hg removal efficiency can be the share of Hg^0 in the flue For AC, CD and CS-400, the Hg removal efficiency decreased with further increase in the adsorption temperature, whereas, for CS-400-Br, RC-600, GRC-600, the efficiency increased

It was speculated that high adsorption temperature causes Hg desorption from the sorbent surface. However, the experiment showed that sorption temperature increase led to a higher share of Hg^0 in the flue gas. The higher Hg removal of CS-400-Br, RC-600 and GRC-600 as the temperature raised can be connected to higher Br and Cl intrinsic content

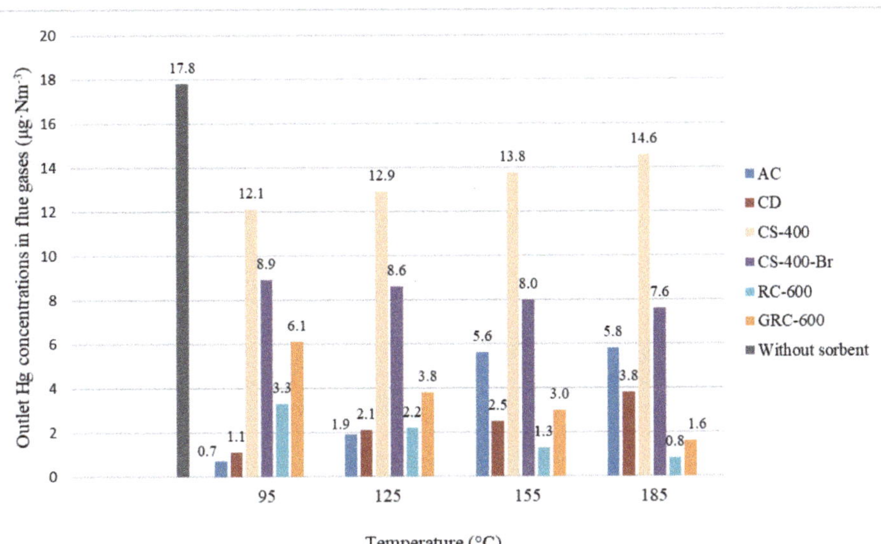

Figure 6. Concentration of mercury in raw flue gas and flue gases cleaned by LCOA depending on the temperature.

Effect of Bromine Content in the Sorbent on Hg^0 Oxidation

Flue gases from lignite combustion are characterized by the high share of unoxidized Hg^0 mercury. It is a volatile constituent of the flue with its ratio versus Hg^{2+} increasing with sorption temperature. As Hg^0 content increases, the average Hg removal efficiency for AC, CD and CS was the highest for 95 °C and the lowest for 185 °C.

AC, CD and CS-400 sorbency was determined by low bromine content, which resulted in lesser Hg^{2+} and therefore higher Hg^0 amounts. The surface of the sorbent allows for Hg^{2+} adsorption, which can explain the pattern of higher sorptive properties of AC, CD and CS-400 at higher sorption temperatures, as the Hg0 content increases.

On the other hand, RC-600, GRC-600 and CS-400-Br were characterized by 2–14 times higher bromine content. Sorbent-present Br oxidized Hg0 in the flue to Hg^{2+} form at higher sorption temperatures, which obtained higher Hg^{2+} concentrations, along with higher removal capabilities [50]. The average Hg removal for RC-600, GRC-600 and CS-400-Br followed the ascending order: 95 °C < 125 °C < 155 °C < 185 °C. The proposed Hg^0 oxidation and Hg^{2+} adsorption mechanism on RC-600, GRC-600 and CS-400-Br surfaces are presented in Figure 7. Hg^0 was first oxidized at the activated site on the sorbent, followed by Hg^{2+} sorption.

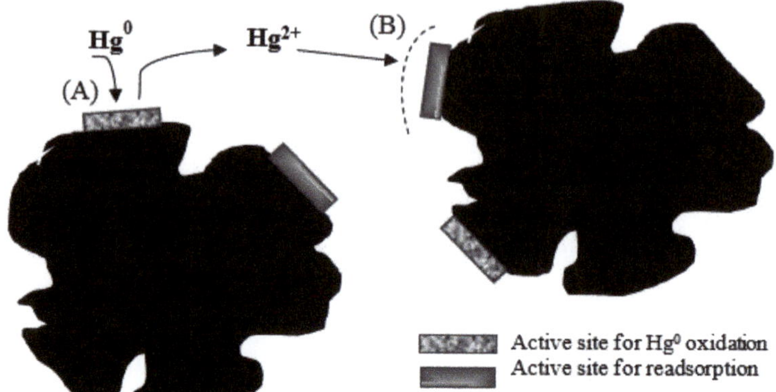

Suggested Hg removal mechanisms of Wa-2 and Sa-4:

(A) Hg^0 oxidation at the activated site.

(B) Readsorption of the resultant oxidized mercury at the site of RC-600/ GRC-600 available for adsorption.

Figure 7. Effect of bromine content in the sorbent on Hg^0 oxidation.

To further assess the application prospect of such rubber waste-derived sorbents, comparison of the accumulative Hg adsorption capacity of RC-600 and GRC-600 with other sorbents was conducted under similar experimental conditions (Figure 8). It demonstrated that longer sorption times (10–60 min, every 10 min) led to increases in the accumulative properties for bromine-present sorbents. No time influence was observed for bromine-free (AC, CD and CS-400) sorbents.

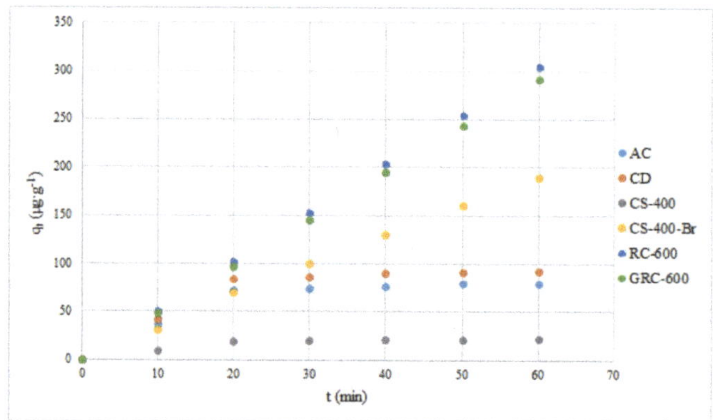

Figure 8. Accumulative Hg adsorption capacity for the tested sorbents.

4. Conclusions

This manuscript presents the results of mercury adsorption from lignite flue gas by the active method. For this purpose, we used six low-cost organic adsorbents (AC, CD, CS-400, CS-400-Br, RC-600 and GRC-600). The presented results allow drawing the following conclusions: the efficiency of sorbents for removal of mercury from flue gases at 95 °C decreased successively: AC (96.1%) and CD (93.8%), followed by RC-600 (81.5%) and GRC-600 (65.7%). The CS-400-Br exhibited better Hg removal performance compared to virgin biochar CS-400. The doping of sorbents with bromine resulted in a higher share of Hg0 oxidation to Hg^{2+} species. Low mercury removal efficiency by CS-400 was caused by its low mesopore volume (0.01 cm^3·g^{-1}). The Hg removal efficiency for AC, CD and biochars (CS-400) decreased with the increase of temperature, caused by a higher amount of sorption inactive Hg^0 mercury in the flue gas.

CS-400-Br, RC-600 and GRC-600 enhanced their mercury adsorption capacity with an increase in the temperature. These sorbents had a higher bromine content (2–14 times) than other sorbents. Therefore, it can be confirmed that bromine had positive effects on Hg^0 removal due to its influence on better Hg^0 oxidation as well as adsorption on the free sites of the surface. In this study, a novel sorbent was created with one-step pyrolysis of tire waste for Hg^0 removal from lignite flue gas. This method could combine municipal solid waste disposal and mercury sorbent preparation in one process. The Hg adsorption capacities of RC-600 and GRC-600 were close to those of commercial activated carbons.

Author Contributions: Conceptualization, M.M.-G. and S.B.; methodology, S.B. and M.M.-G.; validation, K.K.; formal analysis, S.B. and M.M.-G.; investigation, S.B. and M.M.-G.; resources, S.B., M.M.-G. and S.S.; data curation, M.M.-G.; writing—original draft preparation, M.M.-G.; writing—review and editing, M.M.-G., S.B., P.B., B.T. and S.S.; visualization, M.M.-G.; and funding acquisition, M.M.-G. and S.B. All authors have read and agreed to the published version of the manuscript.

Funding: This work was co-financed from the Research Subsidy of the AGH University of Science and Technology for the Faculty of Energy and Fuels (No. 16.16.210.476) and by the National Centre for Research and Development (NCRD) Poland within the LIDER X edition research program. The research and development project is entitled "Prediction of Hg and As distribution during the process of sub-bituminous and lignite coals combustion in pulverised coal-fired boiler and its flue gas cleanup with use of regression models and neural networks" (grant No. LIDER/33/0183/L-10/18/NCBR/2019). This work was supported by research infrastructure of the AGH Center of Energy.

Institutional Review Board Statement: Not applicable.

Informed Consent Statement: Not applicable.

Data Availability Statement: Not applicable.

Conflicts of Interest: The authors declare no conflict of interest.

References

1. Sloane, E.; Ledeboer, A.; Seibert, W.; Coats, B.; Van Strien, M.; Maier, S.F.; Johnson, K.; Chavez, R.; Watkins, L.; Leinwand, L.; et al. Anti-inflammatory cytokine gene therapy decreases sensory and motor dysfunction in experimental Multiple Sclerosis: MOG-EAE behavioral and anatomical symptom treatment with cytokine gene therapy. *Brain Behav. Immun.* **2009**, *23*, 92–100. [CrossRef]
2. Rice, K.M.; Walker, E.M.; Wu, M.; Gillette, C.; Blough, E.R. Environmental Mercury and Its Toxic Effects. *J. Prev. Med. Public Health* **2014**, *47*, 74–83. [CrossRef]
3. U.S. Environmental Protection Agency Office of Air Quality Planning and Standards Air Quality Assessment Division. National Air Quality. Status and Trends through 2007, North. Carolina. 2008. Available online: https://www.epa.gov/sites/production/files/2017-11/documents/trends_brochure_2007.pdf. (accessed on 31 March 2021).
4. Wang, Z.; Liu, J.; Zhang, B.; Yang, Y.; Zhang, Z.; Miao, S. Mechanism of Heterogeneous Mercury Oxidation by HBr over V_2O_5/TiO_2 Catalyst. *Environ. Sci. Technol.* **2016**, *50*, 5398–5404. [CrossRef]
5. Xu, Y.; Ding, H.; Luo, C.; Zheng, Y.; Xu, Y.; Li, X.; Zhang, Z.; Shen, C.; Zhang, L. Effect of lignin, cellulose and hemicellulose on calcium looping behavior of CaO-based sorbents derived from extrusion-spherization method. *Chem. Eng. J.* **2018**, *334*, 2520–2529. [CrossRef]
6. Yang, W.; Liu, Y.; Wang, Q.; Pan, J. Removal of elemental mercury from flue gas using wheat straw chars modified by Mn-Ce mixed oxides with ultrasonic-assisted impregnation. *Chem. Eng. J.* **2017**, *326*, 169–181. [CrossRef]
7. Ren, W.; Duan, L.; Zhu, Z.; Du, W.; An, Z.; Xu, L.; Zhang, C.; Zhuo, Y.; Chen, C. Mercury Transformation and Distribution Across a Polyvinyl Chloride (PVC) Production Line in China. *Environ. Sci. Technol.* **2014**, *48*, 2321–2327. [CrossRef] [PubMed]
8. Zhang, B.; Zeng, X.; Xu, P.; Chen, J.; Xu, Y.; Luo, G.; Xu, M.; Yao, H. Using the Novel Method of Nonthermal Plasma To Add Cl Active Sites on Activated Carbon for Removal of Mercury from Flue Gas. *Environ. Sci. Technol.* **2016**, *50*, 11837–11843. [CrossRef]
9. Li, H.; Zhu, L.; Wang, J.; Li, L.; Shih, K. Development of Nano-Sulfide Sorbent for Efficient Removal of Elemental Mercury from Coal Combustion Fuel Gas. *Environ. Sci. Technol.* **2016**, *50*, 9551–9557. [CrossRef] [PubMed]
10. European Union Emission Inventory Report 1990–2018 under the UNECE Convention on Long-Range Transboundary Air Pollution (LRTAP), Heavy Metal. Emissions, EEA Report No 05/2020. Available online: https://www.eea.europa.eu/publications/european-union-emission-inventory-report-1990-2018 (accessed on 31 March 2021).

11. Guidance on Best Available Techniques and Best Environmental Practices to Control Mercury Emissions from Coal-fired Powe Plants and Coal-fired Industrial Boilers. Available online: https://pdfs.semanticscholar.org/c145/fdc24793cfc6bb614965959753 a585a4b8f.pdf?_ga=2.183433376.761095129.1552896150-1368528823.1551788871 (accessed on 7 December 2020).
12. European Commission. Best Available Techniques (BAT) Conclusions Scope, Ref. Ares 1248230-09/03/2017. Available online https://eur-lex.europa.eu/legal-content/EN/TXT/PDF/?uri=CELEX:32017D1442&from=en (accessed on 31 March 2021).
13. Galbreath, K.C.; Zygarlicke, C.J. Mercury transformations in coal combustion flue gas. *Fuel Process. Technol.* **2000**, *65-66*, 289–310 [CrossRef]
14. Favale, A.; Nakamoto, T.; Kato, Y.; Nagai, Y. Mercury Mitigation Strategy through the Co-Beneift Of Mercury Oxidation with SCF Catalyst. Available online: http://www.mercuryconvention.org/Portals/11/documents/meetings/EG1/Catalyst_HCR.PDF (accessed on 31 March 2021).
15. Liu, Y.; Zhang, J.; Yin, Y. Study on absorption of elemental mercury from flue gas by UV/H2O2: Process parameters and reaction mechanism. *Chem. Eng. J.* **2014**, *249*, 72–78. [CrossRef]
16. Srivastava, R.K.; Hutson, N.; Martin, B.; Princiotta, F.; Staudt, J. Control of Mercury Emissions from Coal-Fired Electric Utility Boilers. *Environ. Sci. Technol.* **2006**, *40*, 1385–1393. [CrossRef]
17. Olson, E.S.; Azenkeng, A.; Laumb, J.D.; Jensen, R.R.; Benson, S.A.; Hoffmann, M.R. New developments in the theory and modeling of mercury oxidation and binding on activated carbons in flue gas. *Fuel Process. Technol.* **2009**, *90*, 1360–1363. [CrossRef]
18. Zhang, J.; Duan, Y.; Zhou, Q.; Zhu, C.; She, M.; Ding, W. Adsorptive removal of gas-phase mercury by oxygen non-thermal plasma modified activated carbon. *Chem. Eng. J.* **2016**, *294*, 281–289. [CrossRef]
19. Zhao, S.; Duan, Y.; Li, C.; Li, Y.; Chen, C.; Liu, M.; Lu, J. Partitioning and Emission of Hazardous Trace Elements in a 100 MW Coal-Fired Power Plant Equipped with Selective Catalytic Reduction, Electrostatic Precipitator, and Wet Flue Gas Desulfurization. *Energy Fuels* **2017**, *31*, 12383–12389. [CrossRef]
20. Charpenteau, C.; Seneviratne, R.; George, A.; Millan, M.; Dugwell, D.R.; Kandiyoti, R. Screening of Low Cost Sorbents for Arsenic and Mercury Capture in Gasification Systems. *Energy Fuels* **2007**, *21*, 2746–2750. [CrossRef]
21. Seneviratne, H.R.; Charpenteau, C.; George, A.; Millan, M.; Dugwell, D.R.; Kandiyoti, R. Ranking Low Cost Sorbents for Mercury Capture from Simulated Flue Gases. *Energy Fuels* **2007**, *21*, 3249–3258. [CrossRef]
22. Liu, Z.; Yang, W.; Xu, W.; Liu, Y. Removal of elemental mercury by bio-chars derived from seaweed impregnated with potassium iodine. *Chem. Eng. J.* **2018**, *339*, 468–478. [CrossRef]
23. Fuente-Cuesta, A.; Lopez-Anton, M.; Diaz-Somoano, M.; Martínez-Tarazona, M. Retention of mercury by low-cost sorbents Influence of flue gas composition and fly ash occurrence. *Chem. Eng. J.* **2012**, *213*, 16–21. [CrossRef]
24. Xu, Y.; Deng, F.; Pang, Q.; He, S.; Xu, Y.; Luo, G.; Yao, H. Development of waste-derived sorbents from biomass and brominated flame retarded plastic for elemental mercury removal from coal-fired flue gas. *Chem. Eng. J.* **2018**, *350*, 911–919. [CrossRef]
25. Marczak, M.; Budzyń, S.; Szczurowski, J.; Kogut, K.; Burmistrz, P. Active methods of mercury removal from flue gases. *Environ Sci. Pollut. Res.* **2018**, *26*, 8383–8392. [CrossRef]
26. Burmistrz, P.; Czepirski, L.; Kogut, K.; Strugała, A. Removing mercury from flue gases: A demo plant based on injecting dusty sorbents. *Chem. Ind. (Przemysł Chem.)* **2014**, *93*, 2014–2019. [CrossRef]
27. Tareq, R.; Akter, N.; Azam, S. *Chapter 10—Biochars and Biochar Composites: Low-Cost Adsorbents for Environmental Remediation, Biochar from Biomass and Waste*; Elsevier Inc.: Amsterdam, The Netherlands, 2019; pp. 169–209. [CrossRef]
28. Szufa, S. Use of superheated steam in the process of biomass torrefaction. *Przem. Chem.* **2020**, *99*, 1797–1801. (In Polish)
29. Szufa, S.; Piersa, P.; Adrian, Ł.; Sielski, J.; Grzesik, M.; Romanowska-Duda, Z.; Piotrowski, K.; Lewandowska, W. Acquisition of Torrefied Biomass from Jerusalem Artichoke Grown in a Closed Circular System Using Biogas Plant Waste. *Molecules* **2020**, *25*, 3862. [CrossRef] [PubMed]
30. Szufa, S.; Wielgosiński, G.; Piersa, P.; Czerwińska, J.; Dzikuć, M.; Adrian, Ł.; Lewandowska, W.; Marczak, M. Torrefaction of Straw from Oats and Maize for Use as a Fuel and Additive to Organic Fertilizers—TGA Analysis, Kinetics as Products for Agricultural Purposes. *Energies* **2020**, *13*, 2064. [CrossRef]
31. Szufa, S.; Piersa, P.; Adrian, Ł.; Czerwińska, J.; Lewandowski, A.; Lewandowska, W.; Sielski, J.; Dzikuć, M.; Wróbel, M.; Jewiarz, M.; et al. Sustainable Drying and Torrefaction Processes of Miscanthus for Use as a Pelletized Solid Biofuel and Biocarbon-Carrier for Fertilizers. *Molecules* **2021**, *26*, 1014. [CrossRef]
32. Li, G.; Shen, B.; Lu, F. The mechanism of sulfur component in pyrolyzed char from waste tire on the elemental mercury removal. *Chem. Eng. J.* **2015**, *273*, 446–454. [CrossRef]
33. Chan, O.S.; Cheung, W.H.; Ckay, G.M. Equilibrium and Kinetics of Lead Adsorption onto Tyre Char. *HKIE Trans.* **2012**, *19*, 20–28. [CrossRef]
34. Selbes, M.; Yilmaz, O.; Khan, A.A.; Karanfil, T. Leaching of DOC, DN, and inorganic constituents from scrap tires. *Chemosphere* **2015**, *139*, 617–623. [CrossRef]
35. Martínez, J.D.; Puy, N.; Murillo, R.; García, T.; Navarro, M.V.; Mastral, A.M. Waste tyre pyrolysis—A review. *Renew. Sustain. Energy Rev.* **2013**, *23*, 179–213. [CrossRef]
36. Manirajah, K.; Sukumaran, S.V.; Abdullah, N.; Razak, H.A.; Ainirazali, N. Evaluation of Low Cost-Activated Carbon Produced from Waste Tyres Pyrolysis for Removal of 2-Chlorophenol. *Bull. Chem. React. Eng. Catal.* **2019**, *14*, 443–449. [CrossRef]
37. Dimpe, K.M.; Ngila, J.C.; Nomngongo, P.N. Application of waste tyre-based activated carbon for the removal of heavy metals in wastewater. *Cogent Eng.* **2017**, *4*, 1–11. [CrossRef]

8. Styszko, K.; Baran, P.; Sekuła, M.; Zarębska, K. Sorption of pharmaceuticals residues from water to char (scrap tires) impregnated with amines. *E3S Web Conf.* **2017**, *14*, 2029. [CrossRef]
9. Gupta, V.; Gupta, B.; Rastogi, A.; Agarwal, S.; Nayak, A. Pesticides removal from waste water by activated carbon prepared from waste rubber tire. *Water Res.* **2011**, *45*, 4047–4055. [CrossRef] [PubMed]
10. Global Mercury Assessment. *Sources, Emissions, Releases and Environmental Transport*; United Nations Environment Programme: Geneva, Switzerland, 2018.
11. ISO/TC 27/SC 5. *ISO 17246:2010, Coal—Proximate Analysis*; International Organization for Standardization (ISO): Geneva, Switzerland, June 2010.
12. ISO/TC 27/SC 5. *ISO 17247:2013, Coal—Ultimate Analysis*; International Organization for Standardization (ISO): Geneva, Switzerland, July 2013.
13. ISO/TC 27/SC 5. *ISO 587:2020, Solid Mineral Fuels—Determination of Chlorine Using Eschka Mixture*; International Organization for Standardization (ISO): Geneva, Switzerland, August 2020.
14. ISO/TC 27/SC 3. *ISO 728:1995—Coke (Nominal Top Size Greater than 20 mm)—Size Analysis by Sieving*; International Organization for Standardization (ISO): Geneva, Switzerland, November 1995.
15. Klobes, P.; Munro, R. *Porosity and Specific Surface Area Measurements for Solid Materials*; National Institute of Standards and Technology: Gaithersburg, MD, USA, 2006. Available online: https://tsapps.nist.gov/publication/get_pdf.cfm?pub_id=854263 (accessed on 13 April 2021).
16. ISO/TC 24/SC 4. *ISO 9277: 2010—Determination of the Specific Surface Area of Solids by Gas Adsorption—BET Method*; International Organization for Standardization (ISO): Geneva, Switzerland, September 2010.
17. ISO/TC 24/SC 4. *ISO 15901-2: 2006—Pore Size Distribution and Porosity of Solid Materials by Mercury Porosimetry and Gas Adsorption—Part. 2: Analysis of Mesopores and Macropores by Gas Adsorption*; International Organization for Standardization (ISO): Geneva, Switzerland, December 2006.
18. ISO/TC 24/SC 4. *ISO 15901-3: 2007—Pore Size Distribution and Porosity of Solid Materials by Mercury Porosimetry and Gas Adsorption—Part. 3: Analysis of Micropores by Gas Adsorption*; International Organization for Standardization (ISO): Geneva, Switzerland, April 2007.
19. ISO/TC 24/SC 4. *ISO 5069-2: 1983 Brown Coals and Lignites—Part 2: Sample Preparation for Determination of Moisture Content and for General Analysis*; International Organization for Standardization (ISO): Geneva, Switzerland, December 1983.
20. Zhang, L.; Zhuo, Y.; Chen, L.; Xu, X.; Chen, C. Mercury emissions from six coal-fired power plants in China. *Fuel Process. Technol.* **2008**, *89*, 1033–1040. [CrossRef]

Article

A Study of Combustion Characteristics of Two Gasoline–Biodiesel Mixtures on RCEM Using Various Fuel Injection Pressures

Ardhika Setiawan [1], Bambang Wahono [1] and Ocktaeck Lim [2,*]

[1]   Graduate School of Mechanical Engineering, University of Ulsan, San 29, Mugeo2-dong, Nam-gu, Ulsan 44610, Korea; ardhika.s@gmail.com or ardhikas@liveuou.kr (A.S.); bambangwahono80@yahoo.co.id or bamb053@lipi.go.id (B.W.)
[2]   School of Mechanical Engineering, University of Ulsan, San 29, Mugeo2-dong, Nam-gu, Ulsan 44610, Korea
*   Correspondence: otlim@ulsan.ac.kr; Tel.: +82-10-7151-8218

Received: 22 May 2020; Accepted: 15 June 2020; Published: 24 June 2020

Abstract: Experimental research was conducted on a rapid compression and expansion machine (RCEM) that has characteristics similar to a gasoline compression ignition (GCI) engine, using two gasoline–biodiesel (GB) blends—10% and 20% volume—with fuel injection pressures varying from 800 to 1400 bar. Biodiesel content lower than GB10 will result in misfires at fuel injection pressures of 800 bar and 1000 bar due to long ignition delays; this is why GB10 was the lowest biodiesel blend used in this experiment. The engine compression ratio was set at 16, with 1000 µs of injection duration and 12.5 degree before top dead center (BTDC). The results show that the GB20 had a shorter ignition delay than the GB10, and that increasing the injection pressure expedited the autoignition. The rate of heat release for both fuel mixes increased with increasing fuel injection pressure, although there was a degradation of heat release rate for the GB20 at the 1400-bar fuel injection rate due to retarded in-cylinder peak pressure at 0.24 degree BTDC. As the ignition delay decreased, the brake thermal efficiency (BTE) decreased and the fuel consumption increased due to the lack of air–fuel mixture homogeneity caused by the short ignition delay. At the fuel injection rate of 800 bar, the GB10 showed the worst efficiency due to the late start of combustion at 3.5 degree after top dead center (ATDC).

Keywords: RCEM; GCI; gasoline; biodiesel; fuel injection pressure

---

1. Introduction

In recent years, economic improvements have increased consumer purchasing power, leading to increased demand for vehicles and electric devices around the world, which will significantly increase the use of fossil fuels. Because fossil fuel sources are finite, many governments have been conducting research to utilize renewable energy as an alternative to the use of fossil fuels. Biodiesel is such an alternative fuel; it can decrease the usage of conventional petroleum diesel over the long term, since it is produced from animal fat and vegetable oil [1].

The compression ignition (CI) engine system is a very promising candidate for this research. It injects fuel near the top dead center (TDC: the position of an engine's piston when it is at the very top of its stroke), which means that only compressed air is used in the compression step. This produces an ideal cycle in the system during combustion, making the performance of the CI engine better than the spark ignition (SI) engine. A high compression ratio can also be applied to CI engines, making them popular in industrial engines and transportation vehicles due to the high energy that can be produced.

In the last two decades, many renewable sources have been popular as energy resources. For the transportation system, alternate fuels—such as vegetable oils, biodiesel and simple alcohol (ethanol and methanol) blends with diesel—have been extensively explored. However, biodiesel has been widely developed as a replacement for diesel fuel. Blends of 20% biodiesel and lower can be used in diesel equipment with no—or only minor—modifications, although certain manufacturers do not extend warranty coverage if equipment is damaged by these blends. The 6% biodiesel to 20% biodiesel blends are covered by the American society for testing and materials (ASTM) D7467 specification. Biodiesel can also be used in its pure form (B100) but may require certain engine modifications to avoid maintenance and performance problems. The Volkswagen Group has released a statement indicating that several of its vehicles are compatible with B5 and B100 made from rapeseed oil and compatible with the European norms (EN) 14214 standard. The use of the specified biodiesel type in its cars will not void any warranty [2]. Mercedes Benz does not allow diesel fuels containing greater than 5% biodiesel (B5) due to concerns about "production shortcomings". Any damages caused by the use of such non-approved fuels will not be covered by the Mercedes-Benz Limited Warranty. The use of biodiesel is not only limited to diesel cars, but is also for airplanes, trains, generators, etc.

However, the diesel fuel that is mainly used for CI produces high emissions, which has become a major environmental concern that has led to the issuance of strict emissions regulations in various countries. Gasoline has lower emissions and has become an alternative fuel to replace diesel. Putrasari et al. [3] found that the thermal efficiency and combustion duration of mixed gasoline–biodiesel is almost the same as diesel. Many researchers have demonstrated that adding biodiesel content to conventional petroleum diesel will decrease pollution to the environment but also decrease the brake power and thermal efficiency because of increased knocking [4]. Gasoline compression ignition (GCI) engines are promising because of their low emissions and high thermal efficiency characteristics, attracting considerable research interest [5,6].

Low-temperature combustion (LTC) can be achieved by using fuel with a low cetane number or high volatility [7,8]. LTC will reduce NOx emissions and improve air–fuel mixing, which lowers the probability of attaining a fuel-rich region and will ultimately reduce particulate matter (PM) [9]. Gasoline is a good choice for LTC, having a high octane number that creates high autoignition resistance. This leads to a longer ignition delay, resulting in a more homogeneous air–fuel mixture. Additionally, the capability of fuel to evaporate depends on its volatility. As a result, enhanced air–fuel mix and decreased local equivalence ratio were obtained.

The low lubricity [10] and higher vapor pressure characteristics of pure gasoline will damage a conventional common rail system over long-term usage [11]. Biodiesel has great potential to overcome the low lubricity problem of gasoline when applied in a GCI. Additionally, given its high oxygen content, adding biodiesel will enhance the combustion process [12]. Adams et al. [13] investigated the combustion behavior of GCI engines, using a mixed fuel of gasoline and biodiesel. They demonstrated that adding biodiesel to conventional petroleum diesel increases the combustion stability and decreases the required intake temperature.

Improving engine performance and efficiency has been actively researched for years. Ahmed et al. [14] studied the effect of a nanomaterial additive on diesel-n-heptane blends in a diesel engine and found an increase in the engine performance. Mehmet [15] investigated the effect of fuel injection pressure on a diesel engine and found that higher fuel injection pressure decreased the brake-specific fuel consumption (BSFC) and enhanced the brake thermal efficiency (BTE) at low speed. Varying the injection pressure seems to be a promising strategy to improve combustion characteristics; it is the main factor that determines fuel stratification inside the chamber and highly affects the combustion process. Higher fuel injection pressure delivers fuel as smaller droplets, resulting in a higher surface area to volume ratio. This enhances the vapor capability of the fuel and produces more complete combustion.

This experiment analyzes the effect of high injection pressure on a GCI engine that uses gasoline as the main fuel. Generally, the GCI engine has an engine structure similar to a diesel engine and uses common rails and injectors with standard diesel fuel. Further study related to high injection pressure using gasoline needs to be done. The purpose of this study was to investigate the combustion characteristics of a GCI engine. A rapid compression expansion machine (RCEM), which has similar characteristics as a CI engine, was used to represent the GCI engine. Gasoline mixed with a small quantity of biodiesel by volume was used with variable high-pressure fuel injections. The purpose of adding biodiesel content to the gasoline was to increase the cetane number to overcome the automatic ignition resistance of gasoline. The injection rate was tested to investigate the effects of the fuel injection pressure on the different densities of gasoline and biodiesel. By varying the injection pressure for the two different fuel mixtures, the combustion characteristics could be analyzed.

2. Methodology

2.1. Fuel Preparation

The two kinds of mixed fuel were investigated: GB10 (90% gasoline and 10% biodiesel) and GB20 (80% gasoline and 20% biodiesel). Biodiesel content lower than GB10 resulted in misfires at fuel injection pressures of 800 bar and 1000 bar due to the long ignition delay. The gasoline–biodiesel mixtures were prepared in glass containers and were stirred for approximately 5 to 15 min to ensure homogeneity. Table 1 shows the chemical composition of biodiesel produced from soybeans [16]. It was necessary to perform our experiments immediately after the mixing process because—given the different densities of the gasoline and the biodiesel—phase separation and crystalline colloid would occur fairly rapidly. Water coolant was added to the fuel containers by means of a stainless steel tube to prevent fuel overheating. The physical properties of the fuels were tested in the laboratory and are shown in Table 2 [3].

Table 1. The chemical composition of soybean oil [16].

Fatty Acid	System Name	Formula	Structure	Composition (wt%)
Erucic	*cis*-13-Docosenoic	$C_{22}H_{42}O_2$	22:1	0
Behenic	Docosanoic	$C_{22}H_{44}O_2$	22:0	0
Linolenic	*cis*-9, *cis*-12, *cis*-15-Octadecatrienoic	$C_{18}H_{30}O_2$	18:3	6
Arachidic	Eicosanoic	$C_{20}H_{40}O_2$	20:0	0
Oleic	*cis*-9-Octadecenoic	$C_{18}H_{34}O_2$	18:1	23
Stearic	Octadecanoic	$C_{18}H_{36}O_2$	18:0	3
Linoleic	*cis*-9, *cis*-12-Octadecadienoic	$C_{18}H_{32}O_2$	18:2	55
Palmitic	Hexadecanoic	$C_{16}H_{32}O_2$	16:0	12
Lignoceric	Tetracosanoic	$C_{24}H_{48}O_2$	24:0	0
Myristic	Tetradecanoic	$C_{14}H_{28}O_2$	14:0	0

Table 2. Physical properties of fuels [3].

Test Item	Unit	Test Method	Gasoline	B100
Heating Value	MJ/kg	ASTM D240:2009	45.86	39.79
Kinematic Viscosity (40 °C)	mm²/s	ISO 3104:2008	0.735	4.229
Lubricity	mm	ISO 12156–1:2012	548	189
Cloud Point	°C	ISO 3015:2008	−57	3
Pour Point	°C	ASTM D6749:2002	−57	1
Density (15 °C)	kg/m³	ISO 12185:2003	712.7	882.3

2.2. Test Engine and System Setup

The experiment was carried out on an RCEM that was designed to replicate the phenomena in one cycle of a CI system. In an RCM/RCEM, the single-shot, rapid compression of a test fuel can be studied in a well-defined and controlled environment without the complex fluid dynamics characteristics of a typical internal combustion engine [17,18]. Figure 1 shows the schematic of the RCEM. It is moved by a 22-kW electric motor and has a 100-mm bore and a 420-mm stroke. Adjusting the screw in the base of the crankshaft will change the compression ratio, which can range from 10 to 23. Temperature sensors were installed in the TDC, body, and BTDC to ensure the uniformity of the initial temperature, which could reach a maximum of 393 K. A Kistler 6052CU20 pressure transducer, connected with a Kistler 5018 amplifier, was used to measure the in-cylinder pressure in the engine. An Autonics E40S8-1800-3-T-24 rotary encoder was used to measure the crank angle position. Both sensors were connected to a Dewetron DEWE-800-CA to log the data. A Bosch 0445110327 seven-hole fuel injector was combined with an injector controller (a Zenobalti ZB-8035 multi-stage injection device) and a common rail solenoid injector peak. To control the injection duration and injection timing, an encoder interface ZB-100 and a ZB-5100 hold driver were used.

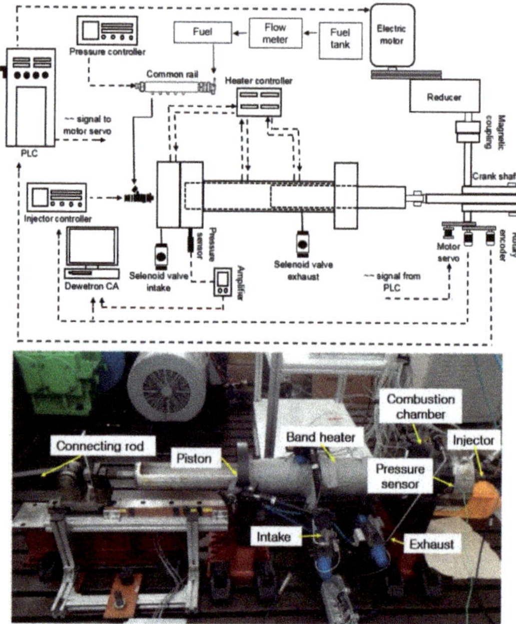

Figure 1. Rapid compression expansion (RCEM) engine setup.

2.3. Test Procedure

The RCEM was able to run up to 1200 RPM, but—given its large bore and stroke—it was considered dangerous to run the engine at such a high speed. A gearbox was installed to reduce the engine speed to 250 RPM, which was maintained as a parameter in this experiment. The fuel injection pressures were varied at 800, 1000, 1200, and 1400 bar, while the fuel injection rate was set at 1000 µs.

To investigate the effect of fuel injection pressure on the different fuel physical characteristics, an injection rate test was conducted. Figure 2 shows the schematic of the fuel measuring system in a glass vessel. A 1.5 bar air pressure was maintained in the vessel by injecting N_2 to create backpressure, as presented in Figure 3. A heating element was used to control the initial temperature within the cylinder and was set at 323 K. Single-fuel injection was controlled at 12.5° before top dead center (BTDC).

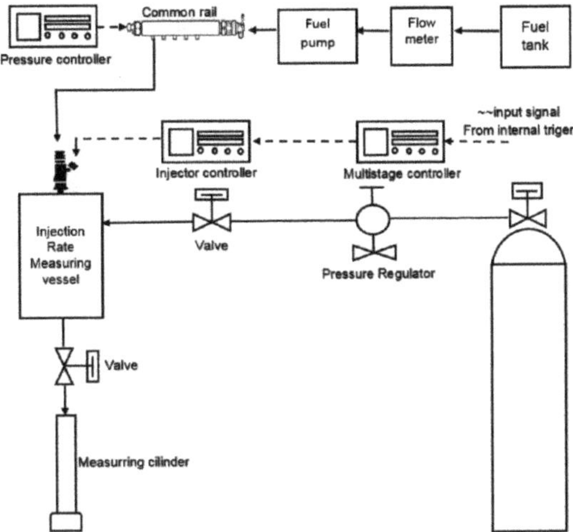

Figure 2. Schematic diagram of the fuel injection test rate measurement system.

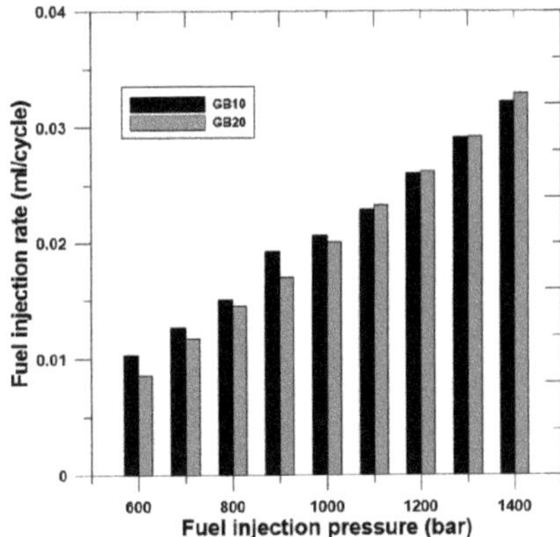

Figure 3. Fuel injection rates of GB10 and GB20.

3. Results and Discussion

3.1. Effect of Fuel Injection Pressure on the Fuel Injection Flow Rate

Figure 3 shows the fuel injection flow rates of the GB10 and GB20 with fuel injection pressure at 600 to 1400 bar. The injection rates between 600 and 1000 bar show clear differences between the GB10 and GB20. The GB10 had higher injection rates than the GB20 due to its lower content of biodiesel, reaching 0.01, 0.013, 0.015, 0.019, and 0.021 mL/cycle, compared to the GB20, which had injection rates of 0.009, 0.011, 0.014, 0.017, and 0.02 mL/cycle, respectively. When the injection pressure exceeded 1000 bar, there were random differences between the GB10 and GB20. The different densities and viscosities of the gasoline and biodiesel caused the different fuel injection rates; a higher-viscosity fuel (i.e., with a higher biodiesel content) will have a lower fuel injection rate. The viscosity affects the fuel injection rate by increasing the viscous friction force and the hydraulic force at the leakage passage. In contrast, the density of the fuel affects only the hydraulic force. We found that higher fuel injection pressure decreased the effects of density and viscosity on the fuel rate, and the difference in injected fuel capacity is similar. These results were in agreement with the previous study by Kim et al. [19], who stated that increasing the fuel injection pressure will decrease the effects of the fuel density because the discharge coefficient becomes steady at a high Reynolds number.

3.2. Effect of Fuel Injection Pressure on Combustion Characteristics

3.2.1. In-Cylinder Pressure and Heat Release Rate

Figure 4 shows the in-cylinder pressures of GB10 and GB20 at different fuel injection pressures. At an 800 bar fuel injection pressure, GB10 produced much lower in-cylinder pressure than the GB20, i.e., 25.78 bar vs. 33.51 bar, respectively. However, GB20 showed less improvement in in-cylinder pressure than the GB10 when a higher fuel injection pressure was applied. At 1400-bar fuel injection, there was only a slight difference of in-cylinder pressure between the GB10 and GB20, i.e., 40.18 bar and 40.22 bar, respectively. From Table 2, we can see that gasoline has a higher heating value than biodiesel. Therefore, fuel with more gasoline is expected to produce greater in-cylinder pressure. However, even with a lower fuel heat value, GB20 at fuel injection pressure 800–1000 bar produces higher in-cylinder pressure than GB10. As shown in Figure 5, the GB10 ignition delay at 800- and 1000-bar fuel injection pressures occurred for a very long time. The start of combustion (SOC) occurs after top dead center (ATDC), after which the pressure in the cylinder decreases due to volume expansion, resulting in reduced in-cylinder pressure generated during combustion. These phenomena were in agreement with Raeie et al. [20], who studied ignition delay by varying the injection timing. At fuel injection pressures of 1200 and 1400 bar, the fuel injection rate of the GB20 was higher than the GB10. 0.026 and 0.0322 mL/cycle, respectively, for the GB10 and 0.0262 and 0.0329 mL/cycle, respectively, for the GB20. This produced higher heating values for the GB20 at those two fuel injection rates.

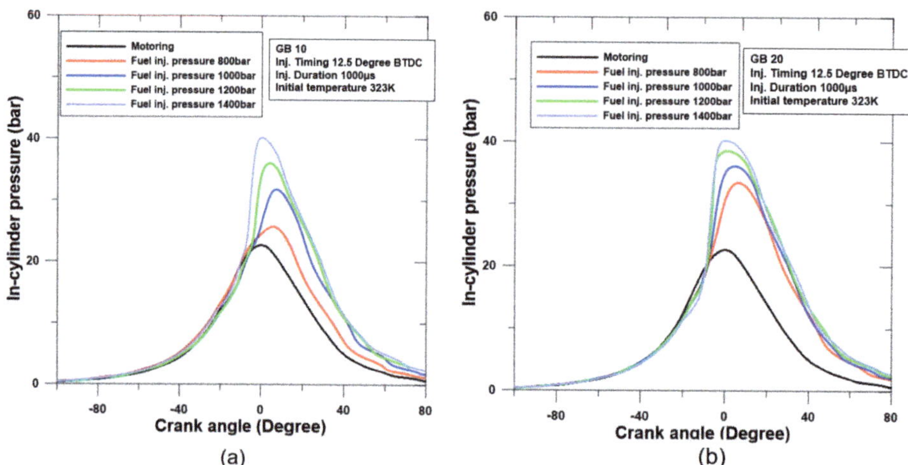

Figure 4. (**a**) In-cylinder pressures of GB10 and (**b**) In-cylinder pressure of GB20.

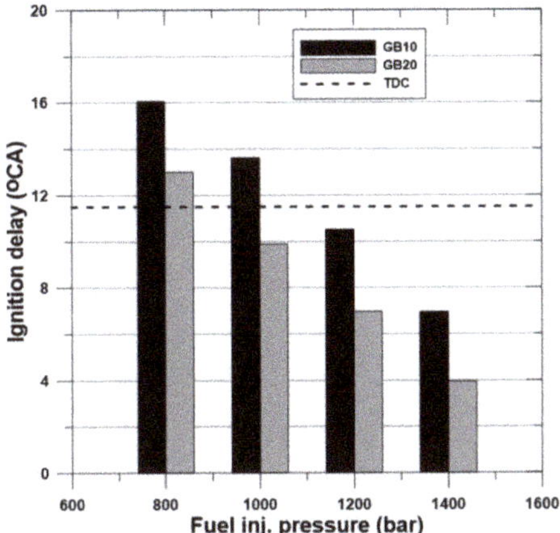

Figure 5. Ignition delays of GB10 and GB20.

As shown in Figure 6, an increase in fuel injection pressure increases the rate of heat release for both fuel mixes, which affects the pressure in the cylinder. With increasing fuel injection pressure, the first stage of heat release or low-temperature heat release (LTHR) also increases. At fuel injection pressures of 1200 and 1400 bar, a significant increase in LTHR is shown for both fuel mixtures. Higher fuel injection pressure will increase spray capability where smaller fuel drops will be produced. This condition will make the fuel more reactive to rising temperatures and pressures, thereby increasing the premixed maximum of HRR. However, at the fuel injection pressure of 1400 bar, the increase in the heat release rate of the GB10 was higher than that of GB20. This was because the early SOC caused in-cylinder peak pressure to occur at 0.24° BTDC, retarding the combustion phasing (CA50) and reducing the heat release rate.

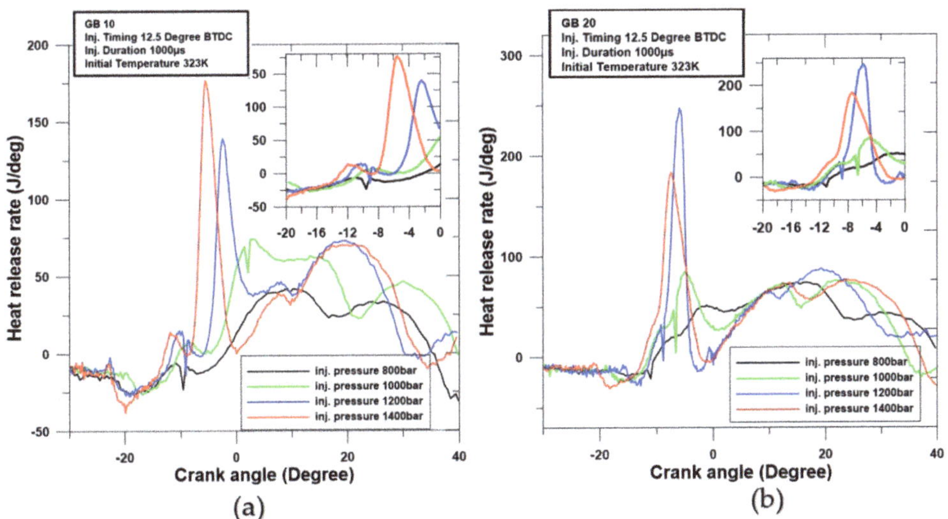

Figure 6. (**a**) Heat release rates of GB10 and (**b**) Heat release rate of GB20.

3.2.2. Combustion Duration and Ignition Delay

The ignition delay of the engine when operated using GB10 and GB20 fuels is presented in Figure 6. In this work, the ignition delay is defined as the time interval of fuel injected until combustion occurs. The second peak of the pressure rise rate is defined as the SOC, as described by Lee et al. [21]. Thus, it can be concluded that the ignition delay will be shortened when higher fuel injection pressure is applied in the engine. Increasing the content of biodiesel also affects the ignition process, making it occur faster. These phenomena are caused by the higher biodiesel content in GB20 than in GB10, which means GB20 has a higher cetane number. Generally, the cetane number is related to the autoignition characteristics of the fuel; the GB20 with its higher biodiesel content and higher cetane number will shorten the ignition delay. This is consistent with Adams [13]. A higher cetane number will also enhance the fuel capability to autoignite against compression, resulting in higher engine performance. Increasing the biodiesel content in a gasoline–biodiesel fuel mix will increase the cetane number, which will improve the autoignition speed and engine performance, as proposed by Putrasari et al. [3]. Meanwhile, an increase in fuel injection pressure increases cavitation—the main breakup—and, consequently, fuel atomization shortens the ignition delay as well as the duration of combustion.

Figure 7 shows the CA10, 50 and 90, while Figure 8 shows combustion duration of GB10 and GB20 with diferent fuel injection pressures. The combustion phasing can be observed in the CA10, CA50, and CA90 graphs. CA50 shows a very consistent advance when higher fuel injection pressure is applied. The start and the end of combustion can be observed from CA10 and CA90, respectively. The combustion duration of GB10 mostly has a lower burn duration than GB20. Fuel with a higher calorific value will produce a shorter total duration of combustion [22]. By increasing fuel injection pressure, shorter combustion duration due to smaller fuel droplets makes fuel more combustible. However, when the rate of increase in the first stage of heat release increases outside the main HRR, the duration of combustion shows an increase. When the first phase of the heat release phase is completed after TDC, the duration of combustion continues to decrease.

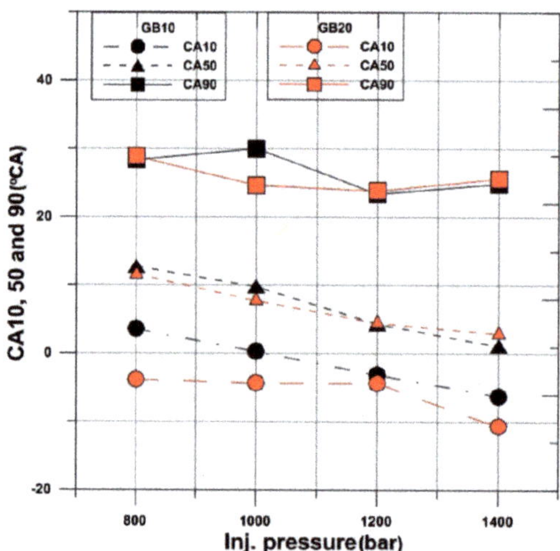

Figure 7. Combustion phasing of CA10, 50 and 90.

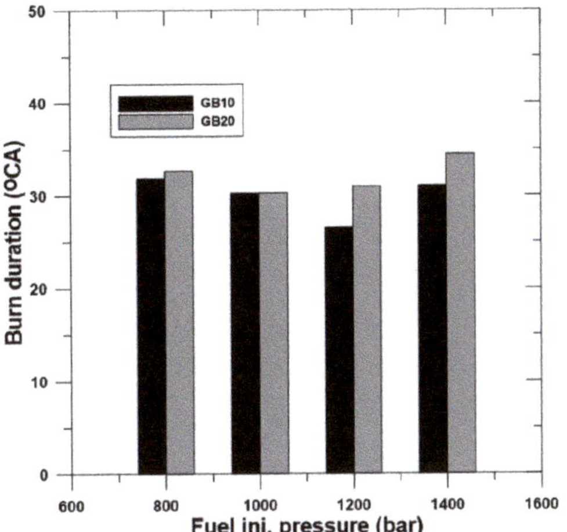

Figure 8. Burn duration.

3.3. Brake Thermal Efficiency and BSFC

Figure 9 shows the brake thermal efficiency (BTE) of the engine using GB10 and GB20 with fuel injection pressures of 800 to 1400 bar. Both fuel blends showed the best improvement of BTE when the SOC occurred near TDC: 31.04% at 1000 bar for GB10 and 55.14% at 800 bar for GB20. Decreases in BTE were shown by both fuel mixes when the fuel injection pressure increased. Figure 10 shows the brake-specific fuel consumption (BSFC) of the engine using GB10 and GB20 with fuel injection pressures of 800 to 1400 bar. GB10 seemed to have a higher BSFC than GB20 in every experimental test: it had the highest fuel consumption, reaching 545 g/kWh at an 800-bar fuel injection pressure, compared to GB20's 360 g/kWh at a 1400-bar fuel injection pressure.

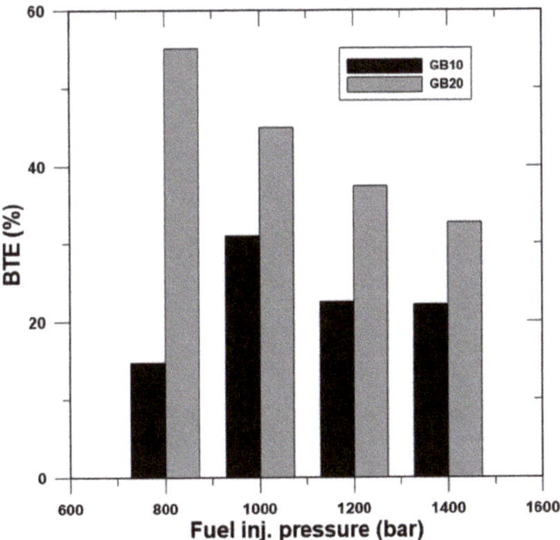

Figure 9. Brake thermal efficiency.

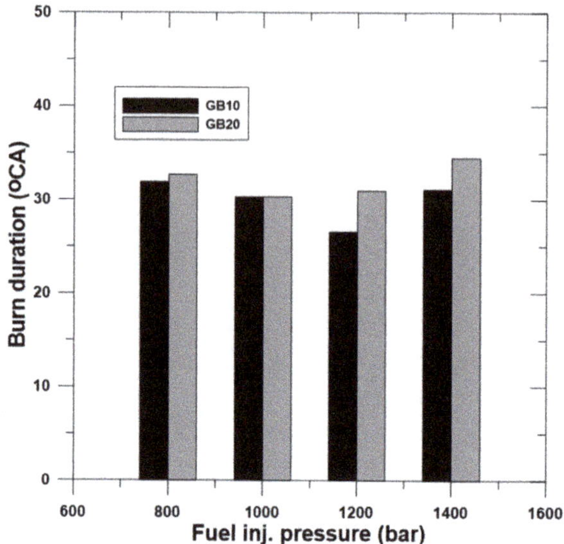

Figure 10. Brake specific fuel consumption.

Increasing the fuel injection pressure will increase the BSFC for all fuel blends. As discussed above, increasing the fuel injection pressure will decrease the ignition delay. A longer ignition delay gives the fuel a chance to mix with the air more perfectly, leading to better quality combustion. However, a too-long or too-short ignition delay will make the engine misfire, and a short ignition delay will produce low combustion quality and efficiency due to insufficient air–fuel mixing. The ignition timing and burn duration will also affect the efficiency of the combustion. However, a too-early SOC will produce backpressure, decreasing the piston speed before it reaches the TDC. A too-late SOC would lead to a decrease in efficiency because of volumetric effects and delayed fuel conversion processes. This statement is in agreement with Haris [23], who studied the effect of ignition delay (ID) on performance, emission, and combustion characteristics of 2-Methyl Furan-Unleaded gasoline blends. He found that retarded ignition will decrease in BTE. For GB10, the SOC at an 800-bar fuel injection pressure occurred after 4.6° ATDC, as shown in Figure 7, resulting in the lowest thermal and fuel efficiency compared with the other samples. The best efficiency was achieved when the SOC occurred synchronously with TDC, which for GB10 was 2.1° CA ATDC at a 1000-bar fuel injection pressure. For GB20, the closest SOC was achieved at an 800-bar fuel injection pressure, which was 2.6° CA BTDC. Meanwhile, by increasing the value of the cetane number, the efficiency also increased. Yakup [24] states that improvement in engine performance by increasing the cetane number may be attributed to improving the combustion process, as the reason for the highest efficiency is shown by GB20 in all tests.

4. Conclusions

This study of the combustion characteristics of two gasoline–biodiesel mixtures using various fuel injection pressures was conducted to gain more understanding of the effects of high fuel injection pressure in a GCI engine. The following conclusions can be drawn from the results of this study.

The injection rates for 600 and 1000 bar show a clear difference between GB10 and GB20, caused by the differences in density and viscosity between gasoline and biodiesel. GB10 results in higher injection rates than GB20 due to the higher content of biodiesel. However, a higher fuel injection pressure will deter the effects of density and viscosity differences on the fuel rate and injected fuel capacity.

If the SOC is closer to the TDC it will maximize combustion, so that it produces high in-cylinder pressure. However, when the SOC exceeds TDC, volumetric inefficiency will cause a pressure drop because GB10 at fuel injection pressures of 800 and 1000 bar has a lower in-cylinder pressure than GB20 while having a higher fuel heat value.

An increase in fuel injection will result in an increase in LTHR. Both fuel mixes showed a significant increase in fuel injection pressure above 1200 bar, which exceeded the main HRR and caused an increase of combustion duration. However, the duration of combustion will still decrease if the end of the first stage of heat release is complete after TDC.

The ignition delay decreased as the fuel injection pressure and the cetane content in the fuel increased. With both GB10 and GB20, increasing the fuel injection pressure above 800 bar produced low efficiency due to the shortened ignition delay. At a fuel injection pressure of 800 bar, GB10 produced the slowest ignition delay or SOC at 3.5° ATDC. At a fuel injection pressure of 1400 bar, GB20 produced the shortest ignition delay, leading to a retarded in-cylinder peak pressure which occurred at 0.24° BTDC, resulting in the lowest efficiency of both fuel mixes.

This study specifically studies the effects of applying high fuel injection pressure in GCI engines with gasoline as the main fuel on combustion characteristics. The idea of increasing efficiency by reducing fuel particles using high fuel injection pressure is not proven to increase efficiency in this system. Not only wasting power, high fuel pressure on gasoline will easily increase the fuel temperature due to fuel friction in the common rail. Fuel with high volatility, such as gasoline, requires a long ignition delay to increase combustion performance. However, further research needs to be done to find out the maximum ignition delay that can be applied to the GCI engine to produce optimal performance.

Author Contributions: Conceptualization, A.S.; methodology, A.S. and B.W.; softwere, A.S.; resource, A.S. and B.W.; writing original draft preparation, A.S.; writing review and editing, A.S.; O.L. supervised the research, advised on the research gap and objective, proofread the paper, and guided the writing process as well as reviewing the presented concepts and outcomes. All authors have read and agreed to the published version of the manuscript.

Acknowledgments: This research was financially supported by the Energy Technology Development Project of the Korea Energy Technology Evaluation and Planning (20182010106370, Demonstration Research Project of Clean Fuel DME Engine for Fine Dust Reduction), Republic of Korea.

Conflicts of Interest: The authors declare no conflict of interest.

References

1.   Kate, J.T.; Teunter, R.; Kusumastuti, R.D.; Van Donk, D.P. Bio-diesel production using mobile processing units: A case in Indonesia. *Agric. Syst.* **2017**, *152*, 121–130. [CrossRef]
2.   Biodiesel Statement. Available online: Volkswagen.co.uk (accessed on 4 August 2011).
3.   Putrasari, Y.; Lim, O. A study on combustion and emission of GCI engines fueled with gasoline-biodiesel blends. *Fuel* **2017**, *189*, 141–154. [CrossRef]
4.   Patel, H.K.; Kumar, S. Experimental analysis on performance of diesel engine using mixture of diesel and bio-diesel as a working fuel with aluminum oxide nanoparticle additive. *Therm. Sci. Eng. Prog.* **2017**, *4*, 252–258. [CrossRef]
5.   Badra, J.; Sim, J.; Elwardany, A.E.; Jaasim, M.; Viollet, Y.; Chang, J.; Amer, A.; Im, H.G. Numerical Simulations of Hollow-Cone Injection and Gasoline Compression Ignition Combustion with Naphtha Fuels. *J. Energy Resour. Technol.* **2016**, *138*, 052202. [CrossRef]
6.   Kodavasal, J.; Kolodziej, C.P.; Ciatti, S.; Som, S. Effects of injection parameters, boost, and swirl ratio on gasoline compression ignition operation at idle and low-load conditions. *Int. J. Engine Res.* **2016**, *18*, 824–836. [CrossRef]
7.   Ra, Y.; Loeper, P.; Reitz, R.D.; Andrie, M.; Krieger, R.; Foster, D.E.; Durrett, R.; Gopalakrishnan, V.; Plazas, A.; Peterson, R.; et al. Study of High Speed Gasoline Direct Injection Compression Ignition (GDICI) Engine Operation in the LTC Regime. *Sae Int. J. Engines* **2011**, *4*, 1412–1430. [CrossRef]
8.   Manente, V.; Zander, C.-G.; Johansson, B.; Tunestål, P.; Cannella, W. An Advanced Internal Combustion Engine Concept for Low Emissions and High Efficiency from Idle to Max Load Using Gasoline Partially Premixed Combustion. *Sae Int. J. Engines* **2010**. [CrossRef]

9. Zheng, M.; Han, X.; Asad, U.; Wang, J. Investigation of butanol-fuelled HCCI combustion on a high efficiency diesel engine. *Energy Convers. Manag.* **2015**, *98*, 215–224. [CrossRef]

10. Hsieh, P.; Bruno, T.J. A perspective on the origin of lubricity in petroleum distillate motor fuels. *Fuel Process. Technol.* **2015**, *129*, 52–60. [CrossRef]

11. Arkoudeas, P.; Karonis, D.; Zannikos, F.; Lois, E. Lubricity assessment of gasoline fuels. *Fuel Process. Technol.* **2014**, *122*, 107–119. [CrossRef]

12. Misra, R.; Murthy, M. Blending of additives with biodiesels to improve the cold flow properties, combustion and emission performance in a compression ignition engine—A review. *Renew. Sustain. Energy Rev.* **2011**, *15*, 2413–2422. [CrossRef]

13. Adams, C.A.; Loeper, P.; Krieger, R.; Andrie, M.; Foster, D.E. Effects of biodiesel–gasoline blends on gasoline direct-injection compression ignition (GCI) combustion. *Fuel* **2013**, *111*, 784–790. [CrossRef]

14. El-Seesy, A.I.; Kosaka, H.; Hassan, H.; Sato, S. Combustion and emission characteristics of a common rail diesel engine and RCEM fueled by n-heptanol-diesel blends and carbon nanomaterial additives. *Energy Convers. Manag.* **2019**, *196*, 370–394. [CrossRef]

15. Şen, M. The effect of the injection pressure on single cylinder diesel engine fueled with propanol–diesel blend. *Fuel* **2019**, *254*, 115617. [CrossRef]

16. Agarwal, A.K. Biofuels (alcohols and biodiesel) applications as fuels for internal combustion engines. *Prog. Energy Combust. Sci.* **2007**, *33*, 233–271. [CrossRef]

17. Goldsborough, S.S.; Hochgreb, S.; Vanhove, G.; Wooldridge, M.S.; Curran, H.J.; Sung, C.-J. Advances in rapid compression machine studies of low-and intermediate-temperature autoignition phenomena. *Prog. Energy Combust. Sci.* **2017**, *63*, 1–78. [CrossRef]

18. Sung, C.-J.; Curran, H.J. Using rapid compression machines for chemical kinetics studies. *Prog. Energy Combust. Sci.* **2014**, *44*, 1–18. [CrossRef]

19. Kim, J.; Lee, J.; Kim, K. Numerical study on the effects of fuel viscosity and density on the injection rate performance of a solenoid diesel injector based on AMESim. *Fuel* **2019**, *256*, 115912. [CrossRef]

20. Raeie, N.; Emami, S.; Sadaghiyani, O.K. Effects of injection timing, before and after top dead center on the propulsion and power in a diesel engine. *Propuls. Power Res.* **2014**, *3*, 59–67. [CrossRef]

21. Lee, S.; Song, S. A rapid compression machine study of hydrogen effects on the ignition delay times of n-butane at low-to-intermediate temperatures. *Fuel* **2020**, *266*, 116895. [CrossRef]

22. Ashok, B.; Nanthagopal, K. Eco friendly biofuels for CI engine applications. In *Advances in Eco-Fuels for a Sustainable Environment*; Elsevier: Amsterdam, The Netherlands, 2019; pp. 407–440.

23. Sivasubramanian, H. Effect of Ignition Delay (ID) on performance, emission and combustion characteristics of 2-Methyl Furan-Unleaded gasoline blends in a MPFI SI engine. *Alex. Eng. J.* **2018**, *57*, 499–507. [CrossRef]

24. Içıngür, Y.; Altıparmak, D.; Altiparmak, D. Effect of fuel cetane number and injection pressure on a DI Diesel engine performance and emissions. *Energy Convers. Manag.* **2003**, *44*, 389–397. [CrossRef]

Article

Comparative Assessment of Spray Behavior, Combustion and Engine Performance of ABE-Biodiesel/Diesel as Fuel in DI Diesel Engine

Sattar Jabbar Murad Algayyim [1,2,3,*] and **Andrew P. Wandel** [2]

1   Department of Mechanical Engineering, University of Al-Qadisiyah, Al-Diwaniyah 58001, Iraq
2   School of Mechanical and Electrical Engineering, University of Southern Queensland, Toowoomba 4350, Australia; andrew.wandel@usq.edu.au
3   Al-Diwaniyah Water Department, Al-Diwaniyah Governorate, Al-Diwaniyah 58001, Iraq
*   Correspondence: sattarjabbarmurad.algayyim@usq.edu.au or ialiraq12@yahoo.com; Tel.: +61-431-291-580

Received: 22 November 2020; Accepted: 7 December 2020; Published: 10 December 2020

Abstract: This study investigates the impact of an acetone-butanol-ethanol (ABE) mixture on spray parameters, engine performance and emission levels of neat cottonseed biodiesel and neat diesel blends. The spray test was carried out using a high-speed camera, and the engine test was conducted on a variable compression diesel engine. Adding an ABE blend can increase the spray penetration of both neat biodiesel and diesel due to the low viscosity and surface tension, thereby enhancing the vaporization rate and combustion efficiency. A maximum in-cylinder pressure value was recorded for the ABE-diesel blend. The brake power (BP) of all ABE blends was slightly reduced due to the low heating values of ABE blends. Exhaust gas temperature (EGT), nitrogen oxides (NO_x) and carbon monoxide (CO) emissions were also reduced with the addition of the ABE blend to neat diesel and biodiesel by 14–17%, 11–13% and 25–54%, respectively, compared to neat diesel. Unburnt hydrocarbon (UHC) emissions were reduced with the addition of ABE to diesel by 13%, while UHC emissions were increased with the addition of ABE to biodiesel blend by 25–34% compared to neat diesel. It can be concluded that the ABE mixture is a good additive blend to neat diesel rather than neat biodiesel for improving diesel properties by using green energy for compression ignition (CI) engines with no or minor modifications.

Keywords: acetone-butanol-ethanol mixture; biodiesel; spray visualization; emissions

1. Introduction

The rapid depletion of fossil fuel reserves, population growth and the increase in air pollution from internal combustion engines using fossil fuels have motivated the search for an alternative biofuel such as biodiesel and alcohol [1,2]. Acetone-butanol-ethanol (ABE), a butanol intermediate product fermentation, has shown potential as an additive fuel blend for conventional diesel due to a reduction in the recovery cost requirements of butanol separation [3–5]. Another benefit of using ABE is that it is produced from renewable sources such as agricultural waste [6–8]. Furthermore, a variety of biomass types can be used as a source of ABE fermentation [3].

The ABE blend has attracted researchers' attention because it is a renewable fuel that reduces dependence on fossil fuels and decreases diesel engine emissions [9,10]. Researchers have experimentally tested ABE mixtures with several investigations [11,12] assessing ABE blend performance under different operating conditions. Luo et al. [13] investigated the sooting tendency of ABE fuels blended with diesel. They found that ABE-diesel has a lower sooting tendency than the butanol-diesel blend because it possesses higher oxygen content and lower carbon content for the same blend ratio.

Ma et al. [14] tested droplet evaporation of an ABE blend. The ABE mixture addition enhanced the droplet evaporation speed, and thus reduced the droplet lifetime. Therefore, droplet clouds have a significant impact on the propagation of turbulent flame. The pre-evaporation rate and droplet size are important parameters in controlling burning velocity. Droplet size and the overall number of droplets also have a substantial impact on ignition success, so the high evaporation rate resulting from the additions of an ABE mixture could improve combustion rates [15]. Recently, a study [16] investigated the impact of a butanol-acetone (BA) mixture as an additive for biodiesel fuel on spray and combustion characteristics. The experimental results revealed that all BA mixtures enhanced spray penetration, offered some improvement in brake power and reduced emission levels (UHC, CO and NO_x). The abovementioned studies support the advantages of using ABE in compression ignition (CI) engines [3,17,18]. Recent research by the authors [5] has also investigated ABE-diesel blends and found that the studied ABE blend reduced exhaust emissions.

To the best of our knowledge, comparative assessment of the ABE mixture as an additive to neat biodiesel (cottonseed) and neat diesel, and the related spray characteristics and engine performance, have not been fully studied. The main goal of this paper is to evaluate and compare the macroscopic spray parameters and engine performance of 10% ABE blended with neat biodiesel and diesel as fuel in a direct injection (DI) diesel engine.

2. Materials and Methods

2.1. Fuel Preparation

Cottonseed biodiesel was prepared from cottonseed oil via transesterification. The fatty acid compositions of the cottonseed biodiesel (chemical profiles) were determined using Flame Ionization Detector-Equipped Gas Chromatography (FID-GC) [19]. The analytical grades of normal butanol (nB) and acetone (A) were used with 99.8% purity, and ethanol was used with 100% purity. All alcohol blends were obtained from Chem-Supply Australia. The ABE blend was mixed with a ratio of A:B:E (3:6:1) by volume and used to simulate ABE fermentation [3]. ABE (10%) was blended with 90% neat cottonseed biodiesel (Bd), which is referred to as ABE10Bd90. ABE (10%) was also blended with 90% neat diesel (D) creating ABE10D90. Blend density was measured according to ASTM 1298 [20,21]. The blends' dynamic viscosities were measured according to the ASTM 445-01 [22] using a Brookfield Viscometer (DV-II+Pro Extra, AMETEK Brookfield, Middleboro, MA, USA). The kinematic viscosity was then calculated [23]. The blends' calorific values were measured using a digital oxygen bomb calorimeter (XRY-1A, Shanghai Changji Geological Instrument Co., Ltd., Shanghai, China) following ASTM D240 [24,25]. The properties of the neat fuel are listed in Table 1, and the measured blends' properties are presented in Figure 1.

Table 1. Fuel properties [3,19,21,24].

Properties	Acetone	N-Butanol	Ethanol	Cottonseed Biodiesel (Bd)	Diesel (D)
Chemical formula	C_3H_6O	C_4H_9OH	C_2H_5OH	-	C_{12}-C_{25}
Composition (C, H, O) (mass%)	62,10.5, 27.5	65,13.5, 21.5	-	9.2, 17.1, 2.9	-
Oxygen content, (mass%)	27.6	21.6	34.78	$\approx$10	0.0
Density (kg/L)	0.971	0.810	0.795	0.864	0.82–0.86
Viscosity (mm^2/s) at 40 (°C)	0.35	2.22	1.08	3.7–4.14	1.9–4.1
Calorific values (MJ/kg)	29.6	33.1	26.8	37.5	42.8
Cetane number	-	17–25	8	52	48
Flash point (°C)	17.8	35	8	128	74
Boiling point (°C)	56.1	118	78.5	280–410	210–235
Latent heat vaporization (kJ/kg)	501.1	582	904	230	270
Auto-ignition temperature (°C)	560	385	434	-	$\approx$300
Surface tension (mN/m)	22.6	24.2	22.27	32.4	23.8
Stoichiometric air–fuel ratio	9.54	11.2	9.02	12.5	15

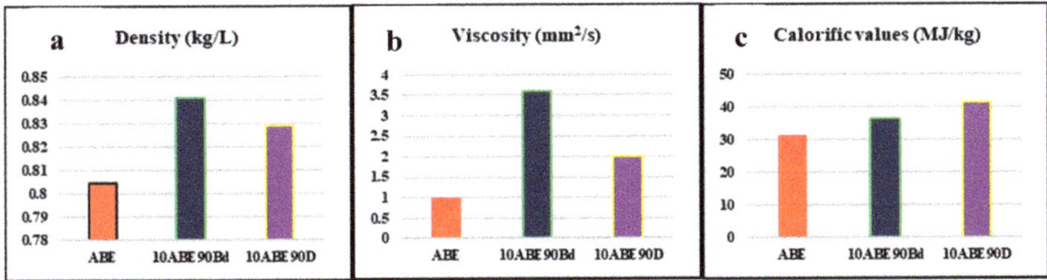

Figure 1. Measured fuel properties. (**a**) Density of test blends; (**b**) Viscosity of test blends; (**c**) Calorifc values of test blends.

2.2. Spray Test Setup

The spray test was conducted at atmospheric condition. The setup was consistent with those described in previous work [4,16,20]. Figure 2 shows a schematic diagram of the spray setup system. The spray images were captured using a Photron SA3 high-speed camera. The injector driver specifications, injection setup and camera specification are presented Table 2. Image processing methods were the same as those employed in [4,16,20,26] using the MATLAB (2015R, The MathWorks, Natick, MA, USA) program.

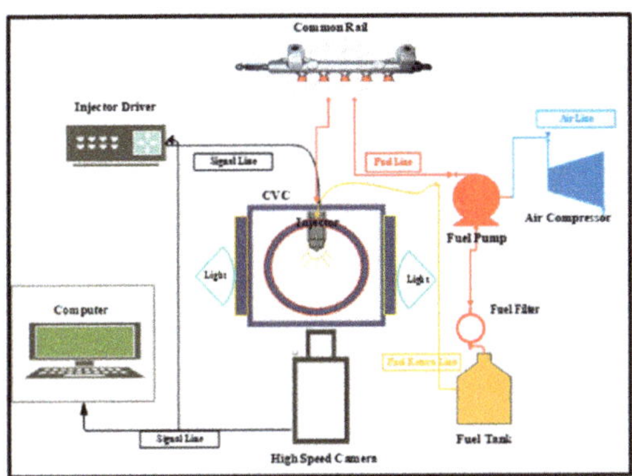

Figure 2. Schematic of spray setup system.

Table 2. Specification of visualization system [4,20].

Injector specification	
Injection type	Bosch electromagnetic common rail injectors solenoid type
Number of nozzles	6 holes
Nozzle diameter (nominal/measured)	0.18 mm.
Camera specification	
Camera resolution @ frame rate	1024 × 1024 pixels @ 2000 fps
A Nikon AF Micro-Nikkor lens with a focal length of 60 mm and a maximum aperture of f/2.8D with filter size 62 mm was connected to the camera	

Table 2. *Cont.*

Injection setup	
Injection Pressure (bar)	300
After start of injection time (ASOI) (mm)	0.5–1.5
Injection enclosed angle (degree)	156
Injection quantity (mg)	12
Repeat time	3

2.3. Engine Test Setup

The engine test consisted of a single-cylinder diesel engine, and a Coda gas analyzer used to measure emissions (Figure 3). Table 3 contains the engine specifications. The engine equipment and the Coda gas analyzer's accuracy ranges were used as described in previous work [16,25]. Specific fuel consumption (SFC) for each fuel test was measured using a flow rate meter because of the differences in density and heating values. Therefore, for each engine run, the test blend properties were included to measure the amount of fuel injected. Brake thermal efficiency (BTE) was calculated from the measured data.

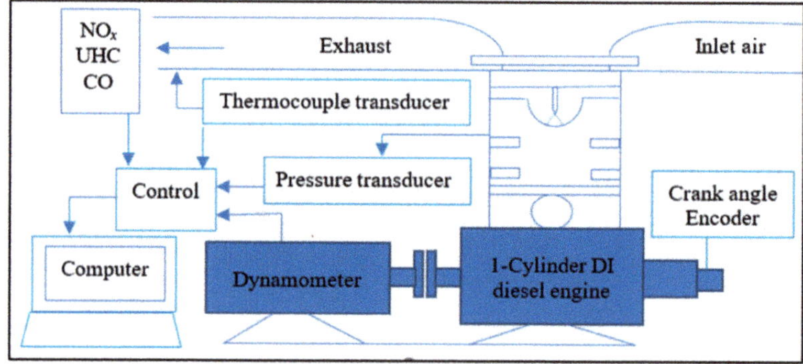

Figure 3. Schematic diagram of test set-up.

Table 3. Engine specifications [5,6].

Engine specifications	
Number of cylinders	1
Compression ratio	5:1–19:1
Bore (mm)	90
Stroke (mm)	74
Capacity (cm^3)	470
Connecting rod	128
Nozzle injection pressure (bar)	300
Nozzle diameter (mm)	0.18
Pressure sensor	Kistler 6052C transducer
Temperature sensor	Thermocouple transducer
Engine test condition	
Engine speeds test @ full load	1400, 2000 and 2600 RPM
Compression ratio test	19:1

3. Results and Discussion

3.1. Spray Characteristics

Spray images of the test fuels are displayed in Figure 4. These images are a sample of three images recorded in the test. The results are presented in Figures 5 and 6. Spray tip penetration (S) and spray cone angle (ϴ) were averaged from three images and six plumes.

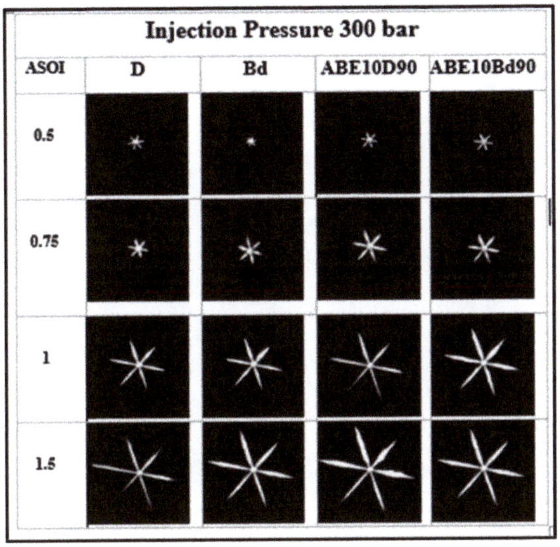

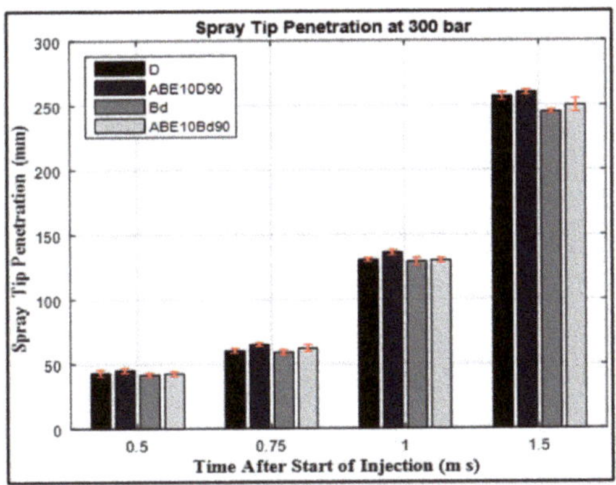

Figure 4. Spray comparison images of test blends.

Spray tip penetration of test fuel blends under different injection conditions is presented in Figure 5 Spray penetration was improved for all ABE test blends because of the low surface tension and viscosity of the ABE blends. The spray penetration of ABE-Bd and ABE-D blends increased, respectively, by 3–5% and 4–5% compared to neat biodiesel and diesel. The low viscosity and boiling point of alcohol can result in improved atomization and evaporation behavior of diesel and biodiesel [26,27]. Therefore, the reaction rate increased. Compared to other test fuels, the spray penetration of biodiesel was clearly shown to be lower due to its high viscosity [26]. Engine power reduction and fuel consumption increments may have occurred because of lower penetration and poorer atomization.

Figure 5. Spray tip penetration of test blends.

The spray cone angle of test blends is presented in Figure 6 under different times after start of injection (ASOI). The increase in injection pressure leads to a slight widening of the spray cone angle. However, the biodiesel fuel presented the maximum spray cone angle due to its high viscosity. In general, at 300 bar injection pressure, the spray cone angle of ABE-D/Bd blends increased at ASOI up to 0.75 ms, while the spray cone angle of neat biodiesel was higher at ASOI 0.5 ms. This result is consistent with the findings of previous work [16,26,27].

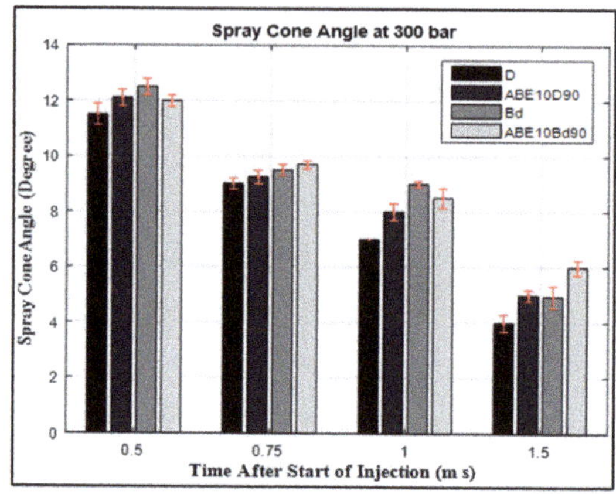

Figure 6. Spray cone angle of test blends.

3.2. Engine Performance

3.2.1. Maximum in-Cylinder Pressure

Figure 7 presents the relationship between the maximum in-cylinder pressure trace of the test fuels. The ABE10D90 blend gave a maximum peak of in-cylinder pressure at 5 bars higher than neat diesel due to the high oxygen content and low cetane number (CN) of the ABE blend. This resulted in increased ignition time and rapid in-cylinder pressure. However, an increase in the engine speed resulted in reduced in-cylinder pressure by about 10 bars. In-cylinder pressure was improved with the addition of ABE to the biodiesel blend. Spray and combustion characteristics enhanced the ABE-biodiesel blend due to a reduction in biodiesel viscosity.

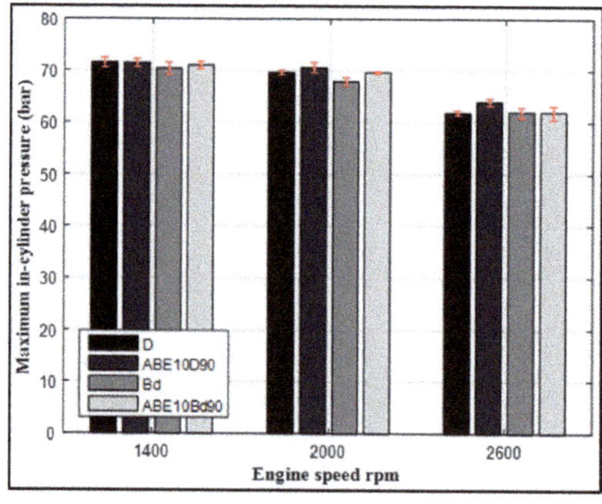

Figure 7. Maximum in-cylinder pressure test fuels.

3.2.2. BP and BTE

The engine was connected to an electrical dynamometer, which was used to measure engine brake power output (BP) at various engine speeds. Brake thermal efficiency (BTE) is the ratio between the brake powers of the engine and the fuel energy supplied to the engine.

Figure 8 shows the variation of BP with engine speed according to test fuel. Both the neat biodiesel (Bd) and ABE-Bd blend revealed a lower value of BP due to low heating values (Table 1 and Figure 1). ABE-D/Bd blends had a higher combustion efficiency because of their high oxygen content, which improved the combustion rate when used as an additive blend. Algayyim et al. [6] investigated the effect of BA-diesel blends in a diesel engine. The experimental results showed that BTE increased because of the addition of BA to the diesel blend. These increments in BTE were achieved because of increased oxygen content in the blend (Figure 9). Oxygen helped to improve combustion efficiency, particularly during the diffusion combustion phase. Another factor influencing the BTE was the cetane number. ABE-diesel/biodiesel fuel blends have a lower cetane number than diesel and biodiesel, causing longer ignition delay, and a wider range in the fraction of fuel burned in the premixed mode, which elevates BTE [23–25].

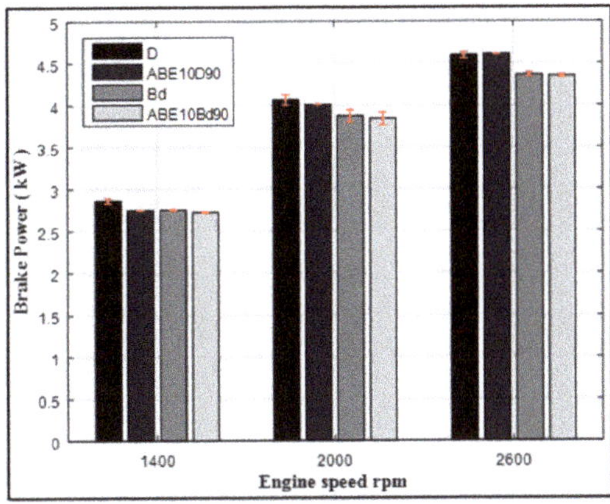

Figure 8. Brake power (BP) of test fuels.

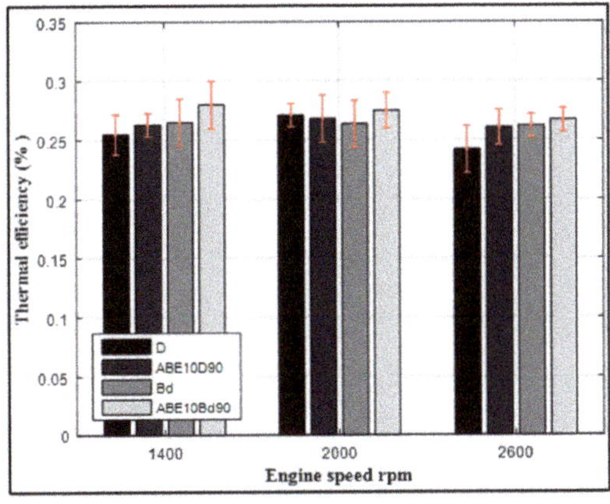

Figure 9. Thermal efficiency of test fuels.

3.2.3. EGT and NO$_x$ Formation

Figures 10 and 11 show the relationship between the EGT and NO$_x$ emissions of the test fuels at various engine speeds. The ABE-Bd blend showed a significant reduction in EGT compared to neat biodiesel at all engine speeds. Neat biodiesel showed higher NO$_x$ emissions compared to conventional diesel (Figure 11) due to the high combustion temperature associated with biodiesel reaction [28–30]. Adding a blend such as acetone, butanol and ethanol creates lower boiling points, which result in an increased evaporation rate and combustion efficiency. Therefore, EGT will be reduced. EGT and NO$_x$ emissions were reduced with the ABE blend to neat diesel and biodiesel by 14–17% and 11–13%, respectively. This result agrees with other work [30–35].

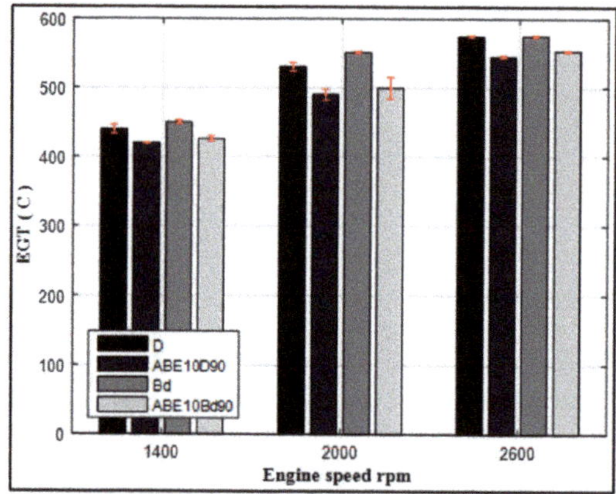

Figure 10. Exhaust gas temperature (EGT) of test fuels.

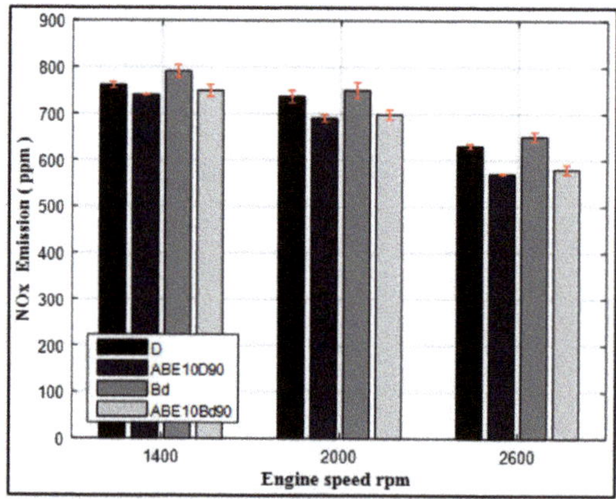

Figure 11. NO$_x$ emissions of test fuels.

3.2.4. UHC, CO and CO$_2$ Emissions

The use of the ABE-D blend decreased UHC emissions compared to neat diesel at medium and high engine speeds (Figure 12). This reduction occurred because the boiling point of the ABE blend was low, which improved the vaporization rate and promoted combustion performance. The difference in droplet lifetime between ABE (3.25 s/mm^2) and diesel (3.75 s/mm^2) affected the reaction time of ABE blends at 823 K, which resulted in increased mixing time and led to complete reaction resulting in decreased UHC emissions [7]. ABE-biodiesel blends increased the UHC emissions due to the high oxygen atom (biodiesel and ABE), which altered the electronic structure. Almost all the C-H bonds of the ABE-Bd blend are less active for reaction compared to hydrocarbon fuels (fossil fuels), which means more time is required to complete the reaction and increases UHC emissions [36,37]. UHC emissions were reduced by 13% when ABE was added to diesel. However, UHC emissions were increased by 25–34% when ABE was added to biodiesel blends. All neat biodiesel and ABE-biodiesel/diesel blends presented lower CO emissions at all engine speeds due to the high oxygen content (Figure 13). This result agrees with previous work [34–37]. Moreover, the oxygen content of ABE-Bd blend was higher than biodiesel resulting in a significant reduction in CO emissions compared to neat biodiesel and ABE-diesel blends. CO_2 emissions were slightly increased of ABE10Bd90 blends (Figure 14). It is clear that ABE0-D and ABE-Bd blends produced less CO emissions compared to diesel due to the lower carbon-to-hydrogen ratio [31,33,37].

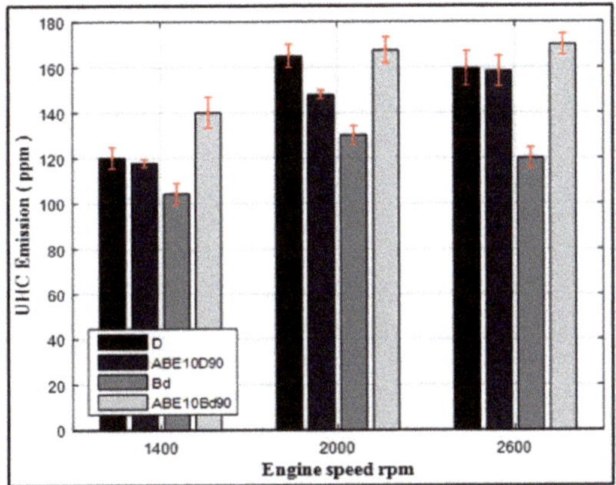

Figure 12. Unburnt hydrocarbon (UHC) emissions of test fuels.

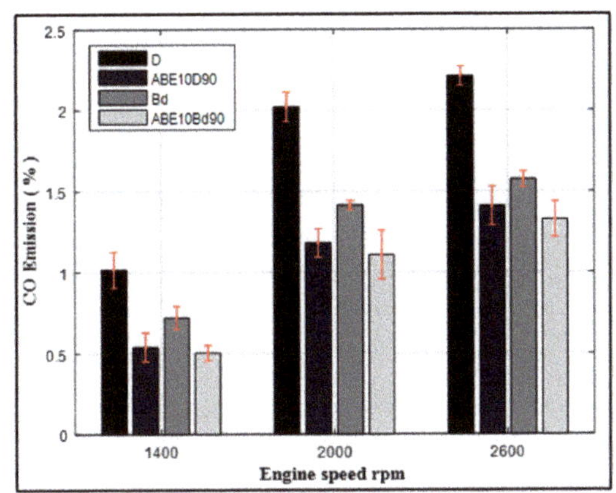

Figure 13. CO emissions of test fuels.

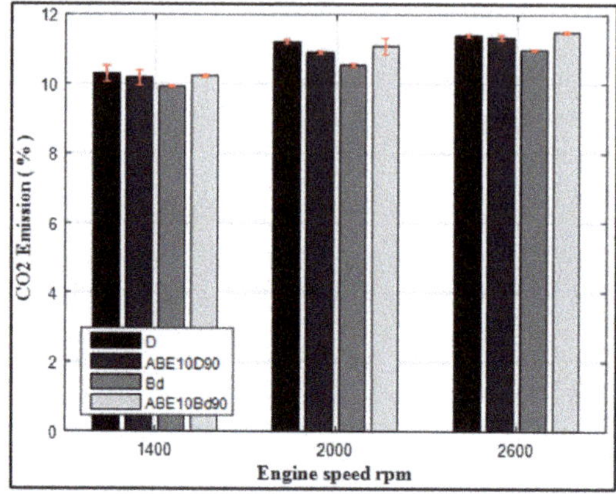

Figure 14. CO_2 emissions of test fuels.

4. Conclusions

ABE mixture can increase the spray penetration of neat diesel and biodiesel resulting in improved atomization. The ABE10D90 blend gives a maximum in-cylinder pressure value 5 bars higher than neat diesel. The Bd and ABE-Bd blends gave a lower value of BP due to the low heating values. EGT, NO_x and CO emissions were significantly reduced with the addition of the ABE blend to both neat diesel and biodiesel. UHC emissions were reduced when ABE was added to diesel. However, UHC emissions were increased when ABE was added to biodiesel. Thus, it can be concluded that the ABE mixture could be a good additive blend for neat diesel rather than neat biodiesel for improving diesel properties by using green energy for CI engines with no or minor modifications.

Author Contributions: S.J.M.A. and A.P.W. conceived and designed the experiments; S.J.M.A. performed the experiments; S.J.M.A. and A.P.W. analyzed the data. S.J.M.A. wrote the paper; A.P.W. provided editorial contribution and guidance. All authors have read and agreed to the published version of the manuscript.

Funding: This research received no external funding.

Acknowledgments: The first author thanks the Iraqi Governments' Ministry of Higher Education and Scientific Research for supporting the research.

Conflicts of Interest: The authors declare no conflict of interest.

References

1. Milazzo, M.F.; Spina, F. The use of the risk assessment in the life cycle assessment framework: Human health impacts of a soy-biodiesel production. *Manag. Environ. Qual.* **2015**, *26*, 389–406. [CrossRef]
2. Nanaki, E.A.; Koroneos, C.J. Comparative LCA of the use of biodiesel, diesel and gasoline for transportation. *J. Clean. Prod.* **2012**, *20*, 14–19. [CrossRef]
3. Algayyim, S.J.M.; Wandel, A.P.; Yusaf, T.; Hamawand, I. Production, and application of ABE as a biofuel. *Renew. Sustain. Energy Rev.* **2018**, *82*, 1195–1214. [CrossRef]
4. Algayyim, S.J.M.; Wandel, A.P.; Yusaf, T.; Hamawand, I. The impact of n-butanol and iso-butanol as components of butanol-acetone (BA) mixture-diesel blend on spray, combustion characteristics, engine performance and emission in direct injection diesel engine. *Energy* **2017**, *140*, 1074–1086. [CrossRef]
5. Algayyim, S.J.M.; Wandel, A.P. Comparative study of engine performance and emissions of butanol, acetone-butanol-ethanol, butanol-acetone/diesel blends. *Biofuels* **2020**. [CrossRef]
6. Algayyim, S.J.M.; Wandel, A.P.; Yusaf, T.; Al-Lwayzy, S.; Hamawand, I. Impact of butanol-acetone mixture as a fuel additive on diesel engine performance and emissions. *Fuel* **2018**, *227*, 118–126. [CrossRef]

7. Lee, T.H.; Lin, Y.; Wu, H.; Meng, L.; Hansen, A.; Lee, C.-F. *Characterization Spray and Combustion Processes of Acetone-Butanol-Ethanol (ABE) in a Constant Volume Chamber*; SAE Technical Paper; SAE: Warrendale, PA, USA, 2015.

8. Lin, Y.; Wu, H.; Nithyanandan, K.; Lee, T.H.; Chia-fon, F.L.; Zhang, C. Investigation of High Percentage Acetone-Butanol-Ethanol (ABE) Blended with Diesel in a Constant Volume Chamber. In Proceedings of the ASME 2014 Internal Combustion Engine Division Fall Technical Conference, New York, NY, USA, 9 December 2014; American Society of Mechanical Engineers: New York, NY, USA, 2014.

9. Zhang, S.; Xu, Z.; Lee, T.; Lin, Y.; Wu, W.; Lee, C.-F. A Semi-Detailed Chemical Kinetic Mechanism of Acetone-Butanol-Ethanol (ABE) and Diesel Blends for Combustion Simulations. *SAE Int. J. Engines* **2016**, *9*, 631–640. [CrossRef]

10. Wu, H.; Nithyanandan, K.; Zhou, N.; Lee, T.H.; Lee, C.-F.F.; Zhang, C. Impacts of acetone on the spray combustion of Acetone–Butanol–Ethanol (ABE)-Diesel blends under low ambient temperature. *Fuel* **2015**, *142*, 109–116. [CrossRef]

11. Wu, H.; Nithyanandan, K.; Lee, T.H.; Lee, C.-F.F.; Zhang, C. Spray and combustion characteristics of neat acetone-butanol-ethanol, n-butanol, and diesel in a constant volume chamber. *Energy Fuels* **2014**, *28*, 6380–6391. [CrossRef]

12. Zhou, N.; Huo, M.; Wu, H.; Nithyanandan, K.; Lee, C.-F.F.; Wang, Q. Low temperature spray combustion of acetone–butanol–ethanol (ABE) and diesel blends. *Appl. Energy* **2014**, *117*, 104–115. [CrossRef]

13. Luo, J.; Zhang, Y.; Zhang, Q.; Liu, J.; Wang, J. Evaluation of sooting tendency of acetone–butanol–ethanol (ABE) fuels blended with diesel fuel. *Fuel* **2017**, *209*, 394–401. [CrossRef]

14. Ma, X.; Zhang, F.; Han, K.; Yang, B.; Song, G. Evaporation characteristics of acetone–butanol–ethanol and diesel blends droplets at high ambient temperatures. *Fuel* **2015**, *160*, 43–49. [CrossRef]

15. Wandel, A.P.; Chakraborty, N.; Mastorakos, E. Direct numerical simulations of turbulent flame expansion in fine sprays. *Proc. Combust. Inst.* **2009**, *32*, 2283–2290. [CrossRef]

16. Algayyim, S.J.M.; Wandel, A.P.; Yusaf, T.; Al-Lwayzy, S. Butanol–acetone mixture blended with cottonseed biodiesel: Spray characteristics evolution, combustion characteristics, engine performance and emission. *Proc. Combust. Inst.* **2019**, *37*, 1–11. [CrossRef]

17. Xu, Z.; Duan, X.; Liu, Y.; Deng, B.; Liu, J. Spray combustion and soot formation characteristics of the acetone-butanol-ethanol/diesel blends under diesel engine-relevant conditions. *Fuel* **2020**, *280*, 118483. [CrossRef]

18. Li, Y.; Tang, W.; Abubakar, S.; Wu, G. Construction of a compact skeletal mechanism for acetone–n–butanol–ethanol (ABE)/diesel blends combustion in engines using a decoupling methodology. *Process. Fuel Technol.* **2020**, *209*, 106526. [CrossRef]

19. Venkatesan, H.; John, J.G.; Sivamani, S. Impact of oxygenated cottonseed biodiesel on combustion, performance, and emission parameters in a direct injection CI engine. *Int. J. Ambient Energy* **2017**, *40*, 1–12. [CrossRef]

20. Algayyim, S.; Wandel, A.P.; Yusaf, T. The Impact of Injector Hole Diameter on Spray Behaviour for Butanol-Diesel Blends. *Energies* **2018**, *11*, 1298. [CrossRef]

21. Delacourt, E.; Desmet, B.; Besson, B. Characterisation of very high-pressure diesel sprays using digital imaging techniques. *Fuel* **2005**, *84*, 859–867. [CrossRef]

22. Yu, W.; Yang, W.; Mohan, B.; Tay, K.L.; Zhao, F. Macroscopic spray characteristics of wide distillation fuel (WDF). *Appl. Energy* **2017**, *185*, 1372–1382. [CrossRef]

23. Al-lwayzy, S.H.; Yusaf, T. Diesel engine performance and exhaust gas emissions using Microalgae Chlorella protothecoides biodiesel. *Renew. Energy* **2017**, *101*, 690–701. [CrossRef]

24. Zaharin, M.S.M.; Abdullah, N.R.; Najafi, G.; Sharudin, H.; Yusaf, T. Effects of physicochemical properties of biodiesel fuel blends with alcohol on diesel engine performance and exhaust emissions: A review. *Renew. Sustain. Energy Rev.* **2017**, *79*, 475–493. [CrossRef]

25. Algayyim, S.J.M.; Wandel, A.P.; Yusaf, T. The effect of Butanol-Acetone Mixture-Cottonseed Biodiesel Blend on Spray Characteristics, Engine Performance and Emissions in Diesel Engine. In Proceedings of the 11th Asia-Pacific Conference on Combustion, New South Welsh, Australia, 10–14 December 2017.

26. Suh, H.K.; Lee, C.S. A review on atomization and exhaust emissions of a biodiesel-fueled compression ignition engine. *Renew. Sustain. Energy Rev.* **2016**, *58*, 1601–1620. [CrossRef]

27. Algayyim, S.J.M.; Wandel, A.P.; Yusaf, T. Experimental and Numerical Investigation of Spray Characteristics of Butanol-Diesel blends. In Proceedings of the 11th Asia-Pacific Conference on Combustion, New South Welsh, Australia, 10–14 December 2017.
28. Xue, J.; Grift, T.E.; Hansen, A.C. Effect of biodiesel on engine performances and emissions. *Renew. Sustain. Energy Rev.* **2011**, *15*, 1098–1116. [CrossRef]
29. Algayyim, S.J.M.; Wandel, A.P. Comparative study of spray characteristics of butanol, acetone-butanol-ethanol, butanol-acetone/diesel. In Proceedings of the Australian Combustion Symposium, Adelaide, Australia, 4–6 December 2019; The University of Adelaide: Adelaide, Australia, 2019.
30. Hajbabaei, M.; Johnson, K.C.; Okamoto, R.A.; Mitchell APullman, M.; Durbin, T.D. Evaluation of the impacts of biodiesel and second-generation biofuels on NOx emissions for CARB diesel fuels. *Environ. Sci. Technol.* **2012**, *46*, 9163–9173. [CrossRef] [PubMed]
31. Song, W.W.; He, K.B.; Wang, J.X.; Wang, X.T.; Shi, X.Y.; Yu, C.; Chen, W.M.; Zheng, L. Emissions of EC, OC, and PAHs from cottonseed oil biodiesel in a heavy-duty diesel engine. *Environ. Sci. Technol.* **2011**, *45*, 6683–6689. [CrossRef] [PubMed]
32. Chauhan, B.S.; Kumar, N.; Cho, H.M.; Lim, H.C. A study on the performance and emission of a diesel engine fueled with Karanja biodiesel and its blends. *Energy* **2013**, *56*, 1–7. [CrossRef]
33. Dhamodaran, G.; Krishnan, R.; Pochareddy, Y.K.; Pyarelal, H.M.; Sivasubramanian, H.; Ganeshram, A.K. A comparative study of combustion, emission, and performance characteristics of rice-bran-, neem-, and cottonseed-oil biodiesels with varying degree of unsaturation. *Fuel* **2017**, *187*, 296–305. [CrossRef]
34. Gopal, K.N.; Ashok, B.; Kumar, K.S.; Raj, R.T.K.; Ashok, S.D.; Varatharajan, V.; Anand, V. Performance analysis and emissions profile of cottonseed oil biodiesel–ethanol blends in a CI engine. *Biofuels* **2017**, *9*, 1–8.
35. Nabi, M.N.; Rahman, M.M.; Akhter, M.S. Biodiesel from cotton seed oil and its effect on engine performance and exhaust emissions. *Appl. Therm. Eng.* **2009**, *29*, 2265–2270. [CrossRef]
36. Al-Muhsen, N.F.O.; Hong, G. *Effect of Spark Timing on Performance and Emissions of a Small Spark Ignition Engine with Dual Ethanol Fuel Injection*; SAE International: Warrendale, PA, USA, 2017.
37. Algayyim, S.J.M.; Wandel, A.P.; Yusaf, T. Mixtures of n-butanol and iso-butanol blended with diesel: Experimental investigation of combustion characteristics, engine performance and emission levels in CI engine. *Biofuels* **2018**, 1–10. [CrossRef]

Publisher's Note: MDPI stays neutral with regard to jurisdictional claims in published maps and institutional affiliations.

Article

Identification of the Effects of Fire-Wave Propagation through the Power Unit's Boiler Island

Michał Paduchowicz * and Artur Górski

Department of Machine and Vehicle Design and Research, Faculty of Mechanical Engineering,
Wroclaw University of Science and Technology, 27 Wybrzeże Wyspiańskiego, 50-370 Wrocław, Poland;
artur.gorski@pwr.edu.pl
* Correspondence: michal.paduchowicz@pwr.edu.pl; Tel.: +48-71-320-39-08

Abstract: The article presents the results obtained during the inspection of the load-bearing structure of a power unit that suffered from fire. The inspection, consisting in the assessment of both the structure's technical condition and durability of welded joints, was performed on seven height levels of the power unit. The vibration spectrum of the unit's steel structure was analyzed, and frequency characteristics were, thus, obtained for individual measurement levels. Thermal vision measurements were also performed in the unit's all connection points to check for possible unsealing of some elements in the boiler island of the inspected power unit. The next stage consisted of performing strength calculations of the steel structure with a goal to estimate the structure's stress state. The conclusions contain suggestions for modernization of welded joints in order to maintain the power unit's design strength.

Keywords: power boilers; the load-bearing structures; damage assessment; fires; tanks

Citation: Paduchowicz, M.; Górski, A. Identification of the Effects of Fire-Wave Propagation through the Power Unit's Boiler Island. *Energies* **2021**, *14*, 1231. https://doi.org/10.3390/en14051231

Received: 2 December 2020
Accepted: 10 February 2021
Published: 24 February 2021

Publisher's Note: MDPI stays neutral with regard to jurisdictional claims in published maps and institutional affiliations.

1. Introduction

The article addresses the problem presented in many other publications on the spread of fires in industrial installations [1]. As the results of the article [2] show, the main causes of fires in such places as coal-fired power plants, where hard coal are used, are explosive works, self-heating as well as friction and impact. In this study, an attempt was made to verify the causes of the fire and to verify the effects of the passage of the fire wave through the area of power units covered with brown coal together with biomass. The thesis was put forward that the combustion of these two fuels is dangerous at the same time due to the self-heating of biomass, which may lead to its spontaneous ignition. This, in turn, taking into account the ease and speed of coal dust (brown coal is crushed into a dust fraction and dried at the same time before being fed to the combustion chamber), may affect the intensity of combustion, which in turn may cause an explosion at high concentration of this element. If we take into account the fact that in the vicinity of fluidized bed power units, there are large amounts of coal dust (among others on communication elements such as platforms, barriers), this will largely help to spread the fire wave within the boiler installations.

Within a power unit's boiler island, a passing fire wave affected the unit's second pass (convection chamber), the cyclone, ash removal system, air-supply system, etc. The point of origin and the path of the fire wave are shown in Figure 1 [3]. References should be numbered in order of appearance and indicated by a numeral or numerals in square brackets, e.g., see the end of the document for further details on references [4,5].

The fire was initiated in the conveyor belt gallery situated in the coal and biomass supply line of power unit A. The fire wave later traveled through communication lines located between the power units and approached further generating units. The propagation of the fire wave necessitated evaluation of the degradation and damage caused to particular elements of the power units. The tests and measurements in this case were complex and consisted of [6]:

89

-   Geometric measurements of the power unit's main components: load-bearing structure, second pass chamber, cyclone, ash removal system, air ducts, accessory elements, etc.;
-   Identification of deviations from technical documentation and measurement results prior to the fire;
-   Non-destructive measurements of accessible welded joints in the power unit's main components, load-bearing structure, second pass convection chamber, cyclone, ash removal system, air ducts, and auxiliary elements, etc., in order to find potential material discontinuities;
-   Evaluation of damage levels for selected spots of the investigated power unit's elements after the fire;
-   Weld quality tests (continuity tests) regarding geometric welding imperfections [7];
-   Measurement of vibration levels for the power unit's load-bearing structure in order to identify vibration amplitude;
-   Thermal imaging measurements of temperature distribution on particular elements of the inspected power unit;
-   Noise level measurements, in particular, selected spots of the power unit;
-   Development of necessary selected computational models for the power unit's components that allow for computer simulations of the impact of fire wave on the structure's shape. Numerical analysis of the likelihood that load-bearing structure, furnace chamber and second pass, cyclone, ash removal system, air ducts, auxiliary elements, etc. suffered from permanent deformations caused by fire wave.

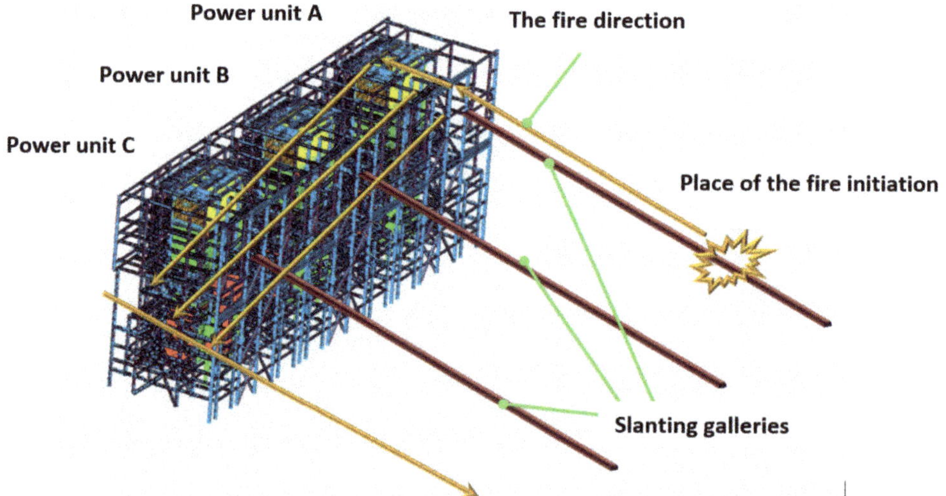

Figure 1. The fire's point of origin and way through the power units.

Determination of the actual state of effort of structural elements subjected to thermal loads is crucial in the assessment of the technical condition of power unit elements. The literature presents many practical methods; however, in the case of such a complex structure as a power unit with the entire installation of water, steam, coal, lime and air, its own methodology was proposed, consisting in the integration of numerical and experimental methods. The main assumption of the methodology is the dependence of the value of reduced stresses on temperature (1).

$$\sigma_i^{(T)} = -\frac{E\beta(T - T_0)}{1 - 2\nu} \tag{1}$$

where the temperature field is described (2):

$$\frac{\partial}{\partial x_i}\left[\lambda_{ij}\frac{\partial T}{\partial X_j}\right] + q_V = \rho c \frac{\partial T}{\partial t} \tag{2}$$

- T—medium temperature;
- λ_{ij}—matrix of thermal conductivity coefficients;
- ρ—density;
- q_v—heat generated in the medium;
- c—pecific heat;
- t—time.

The research on the causes of the fire was based on simultaneous thermal and structural analyses, which are presented in the diagram below (Figure 2).

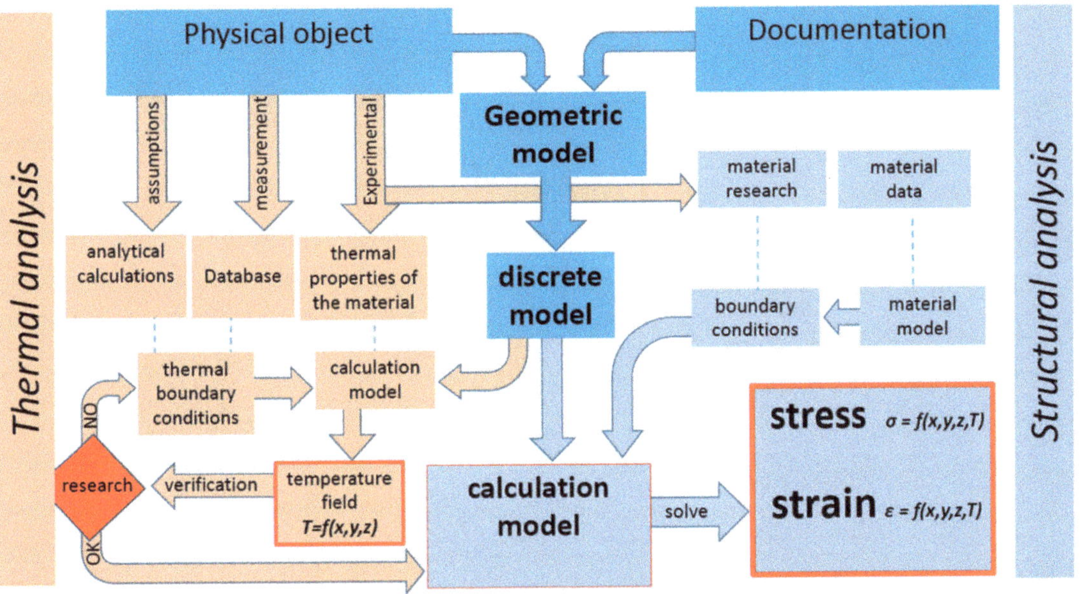

Figure 2. Scheme of tests carried out on the power unit where the fire occurred.

The research scheme presented in the figure above refers to the numerical analysis performed using the finite element method. First, the thermal analysis of the power device model is carried out, and then, using the obtained spatial temperature field and mechanical loads, the complex state of stress of this structure is determined.

2. Operating Parameters of Fluidized Bed Power Unit

For the past few years, electric energy producers have been obliged by EU legal regulations to cofire biomass with coal in conventional power units. This requirement compromises fire safety within the area of energy facilities in operation. One of the greatest threats is coal autoignition resulting from its coming into contact with biomass of typically high temperature. The above case was the hypothetical reason behind the ignition of fire in one of the biomass supply galleries in the analyzed power unit [8].

Apart from that, the use of biomass as one of the ingredients included in the fuel supplied to the boilers in conventional power units greatly influences operating conditions of the whole boiler system. The most frequent problem consists of lowered thermal efficiency, as biomass is less energetic than the fuel originally intended in the power unit's design. This is mainly due to the creation of residue on heat exchange surfaces and the start of corrosion and erosion processes on those elements. Fuel with biomass admixture is usually burnt in fluidized bed boilers. Figure 3 shows a schematic diagram of such type of power unit [9].

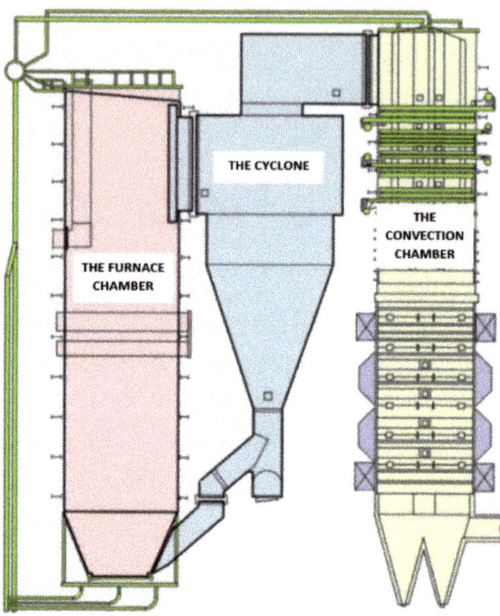

Figure 3. A scheme of fluidized bed boiler unit.

Table 1, on the other hand, shows technical parameters of a fluidized bed boiler unit.

Table 1. Operating parameters of CFB-670 boiler.

Parameter	Value
Air excess factor l	1.1–1.2
Temperature in the bed, Tz	850–870 °C
Burnup fraction	96–98%
Sulfur fixation efficiency	90%
NOx emission (NOx content in flue gas)	$(Ca/S = 1.5–2.6)$
	200–400 mg/m^3

3. Non-Destructive Tests

The steel structure of the power units that experienced the fire was inspected for damage. The inspections included inter alia welded joints of the unit's main components, i.e., load-bearing structure, casing of both furnace and convection chambers, cyclone, ash removal system, air ducts, and auxiliary elements. The objective was to find potential material discontinuities (fractures) and to evaluate degradation level of selected areas on boiler elements against the required standards. Where necessary (and where possible), non-destructive tests were performed by means of magnetic particle inspection with permanent magnet magnetization. In order to detect magnetic stray field, during the magnetization process, the inspected element was sprayed with magnetic particles. In order to obtain clearer MPI magnetograms, the inspected areas were degreased and painted with undercoater. The first stage consisted of a visual inspection of those external surfaces that had close contact with the fire wave. Sample results are shown in Figures 4 and 5

Special scrutiny was given to welded joint and bolted joint areas in the main members of the boiler's support structure [10].

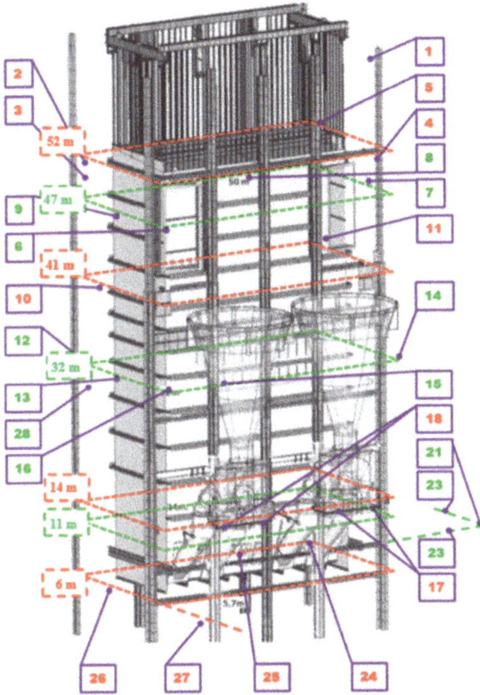

Figure 4. Diagram of the power unit showing the place and labels for inconsistencies found.

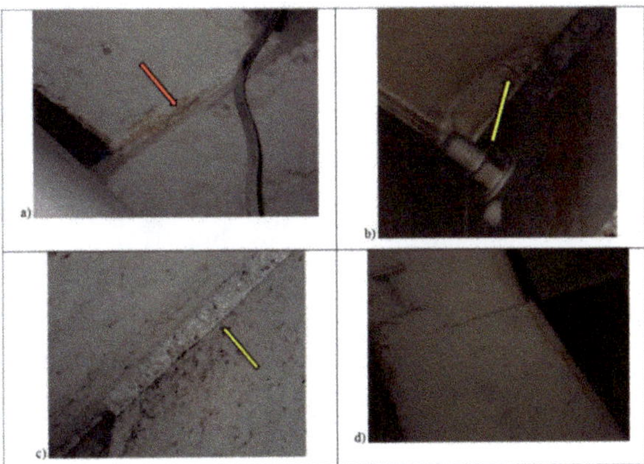

Figure 5. Weld cracks in the girders of the unit's load-bearing structure: (**a**) weld discontinuity between the face and the heat-affected zone (HAZ); (**b**) weld crack; (**c**) subsurface crack of the girder's weld; (**d**) weld crack at the connection point between load-bearing girders.

The inspected areas of the power unit were shown in Figure 4 and marked with schematic labels.

Examination of the power unit's load-bearing structure was mainly focused on the strength of load-bearing members, their connections, and joints. Inspections of load-bearing structure were performed on seven height levels (6, 11, 12, 32, 41, 47, and 52 m) indicated in Figure 4. The inspections covered welded joints and weld quality, as well as bolted joints and plastic strain of both beams and casing. The obtained results allowed to make the following observations [11]:

-   Cracks of weld line in many places (Figure 5);
-   Insufficient quality of welding and weld repairs, welded joints made in inappropriate places;
-   Elastic buckling of lattice work in the load bearing structure;
-   Repairs done during power unit's operation were not protected against corrosion;
-   Inappropriate bracing, which may cause stress concentration in the areas of their construction;
-   Additionally, a great number of welded joints is accumulated on a small area, which leads to the change in material structure, as a lot of heat is supplied and, therefore the material loses its strength characteristics;
-   Deplanation (warping) of flanges and webs in support beams, as well as deformation of reinforcing ribs in I-beam girders;
-   Forbidden process holes (mount holes) in load-bearing girders and posts.

In order to estimate the effects of fire wave propagation from coal and biomass feeders to the boiler, control was needed of the degradation level of the boiler's load bearing structure and of all machines within the analyzed power unit. To meet that goal inspections of the boiler's systems and steel structure were performed. Inspection of the steel structure revealed faults that took the form of cracks in base material and between the weld and the heat affected zone, as shown in Figure 5. Much part of the structure's damage was caused by high temperature, which significantly lowered the strength of load-carrying elements to standard loads. That is when loss of stability took place. The inspection also revealed defects of the structure, caused by incorrectly performed assembly procedures and resulting in the low quality of welded joints, as well as in forbidden notches, cuts, and mount holes or process holes. The above-listed faults contributed to significantly decreased strength of the discussed elements. Additionally, geometrical notches cut in the structure (structural notches) significantly contributed to lowered durability of those elements. Apart from the above, the inspection also allowed to observe numerous signs of repairs, which had been performed during operation of the power unit and were never protected against corrosion, resulting in the appearance of numerous corrosion centers. During inspection attention was also paid to inappropriate bracings, which may cause stress concentration in the areas of their construction. Incorrectly mounted, additional metal sheets create a structural notch, which means rapid change in stiffness, resulting in the interruption of proper load distribution. In addition, a great number of welded joints is accumulated on a small geometric area, which leads to the change in the material's structure due to great amount of heat supplied during welding. This causes the structural material to lose its strength [12,13].

4. Measurement of Vibration Levels in the Load-Bearing Structure of the Power Units

The planned scope of research also included measurements of vibrations that occur in the combustion chamber's load-bearing (support) structure, on its individual height levels, between 5 and 50 m. Figure 6 shows a general diagram of measurement points arrangement for the 32-m level [14].

Measurements of vibration levels were performed in two perpendicular directions on the load-carrying posts of the power unit's separate height levels. The diagram of separate measurement points, as well as the directions of the analyzed displacement and acceleration spectra are shown in Figure 7. The measurements allowed to obtain results in the form of vibration frequency characteristics for individual measurement points. Sample measurement results are shown in the following figures. Values of maximum vibration amplitudes (A) for the measured time range are given below the illustrations [14].

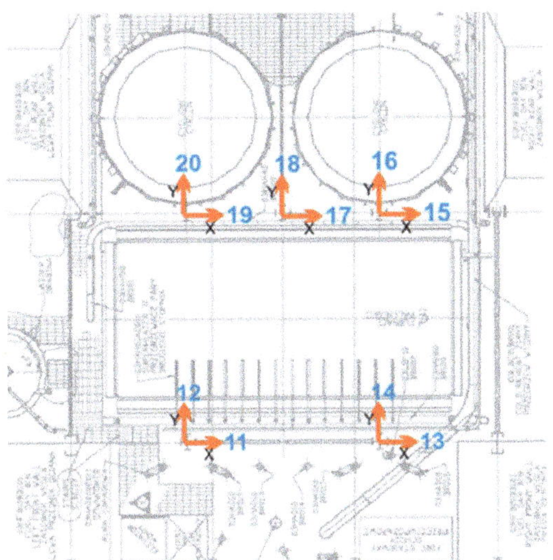

Figure 6. Diagram of measurement points on units 1, 2, and 3—the 32-m level.

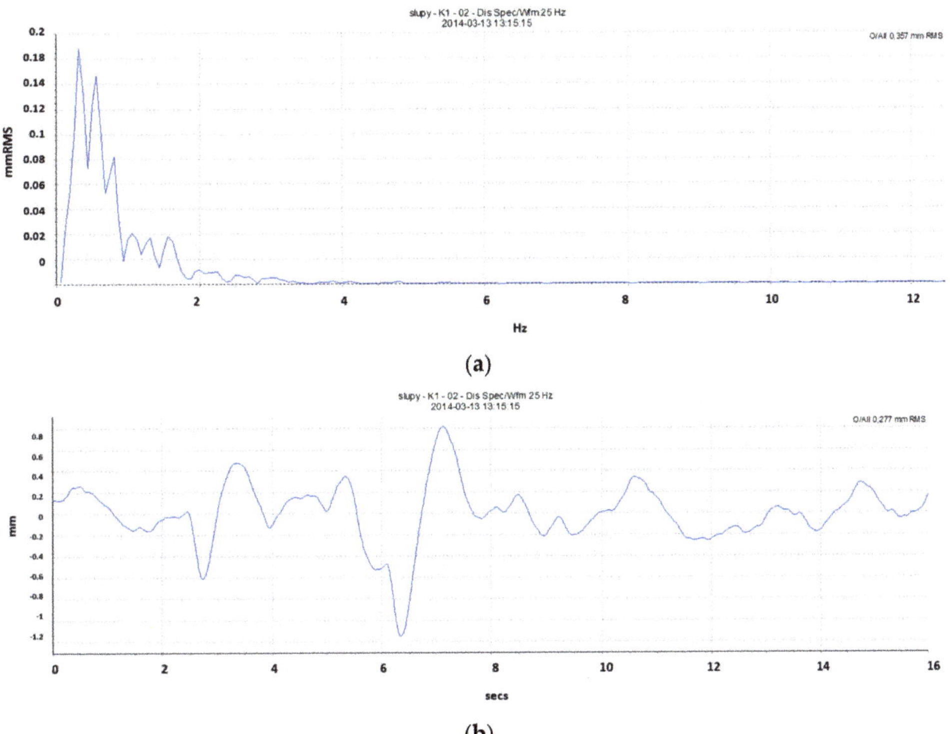

Figure 7. Diagrams of accelerations and displacements in a sample point 2, where vibration amplitudes for the measuring range are inter alia (**a**) 0.815 ÷ 0.191 mmRMS, (**b**) −1.146 ÷ 1.961 mm.

5. Measurements of Temperature Distribution on Particular Elements of the Power Unit

The measurements covered heat loads of the whole power unit, including its load-bearing structure and other selected process equipment of the boiler island, and were performed using a thermal imaging camera with adequate measuring range and sensitivity. The temperature distribution, thus, obtained was used in further analysis, as one of the boundary conditions for structural strength calculations based on finite element method (FEM). Sample results of the measurements performed for the boiler's furnace chamber and auxiliary equipment are shown in Figures 8 and 9 [15].

Figure 8. Temperature distribution at the soot blower entrance point into the furnace chamber.

Figure 9. View of temperature distribution on the external casing of flue gas cyclone—clearly visible thermal bridges in the cyclone's upper part.

Thermal imaging inspection of temperature distribution was performed for the power unit's height levels in the range between 0 and 56 m. In addition, the obtained thermal images allowed us to estimate the quality of thermal insulation used in the furnace chamber's external casing and other boiler equipment. This allowed us to find areas with damaged thermal insulation and the resulting thermal bridges, especially those located on the connecting points of the machines working with furnace chamber and other elements. The damage of external insulation layers was closely connected to the propagation of the fire wave in the vicinity of the power units [16].

6. Strength Calculations for the Load-Bearing Structure of the Power Unit and Its Elements

The measured temperature distribution was used to perform FEM strength calculations for some elements of the power units' parts, such as their load-bearing structure, cyclone, and fuel hopper. The above-listed elements were particularly subjected to heat loads resulting from the fire that occurred in the boiler hall, as they do not have insulated external envelope. The calculations were performed using the finite element method in the NX-IDEAS system [17]. For the sake of the calculations, geometric models of the abovementioned elements were prepared (Figure 10), which were later digitized using appropriate finite elements. The next step consisted in strength calculations, with consideration paid to local temperature changes on external surfaces of the elements [18].

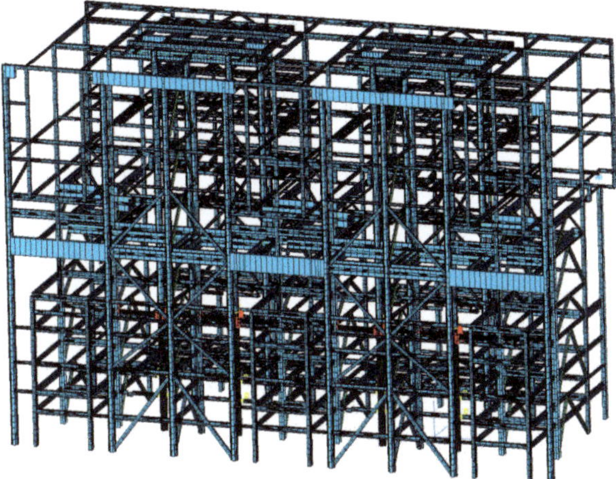

Figure 10. Load-bearing structure of the power unit—general view.

The calculations included both thermal and mechanical loads simultaneously. These are both gradients of temperature, which occur in steel structure due to its uneven heating resulting from the fire, and of structural loads, which occur due to the weight of subassemblies hung on the boiler island's steel structure. However, before the structure's effort was calculated, the spatial thermal field was found for the power unit's support structure. This kind of distribution is highly likely to occur in real structures. It is shown in Figure 11. The results of strength calculations obtained in the form of general displacement contour and Huber–Mises reduced stress contour are shown in Figure 12 [19].

Figure 11. Thermal field contour lines (°C) in the whole load-bearing structure of the power unit.

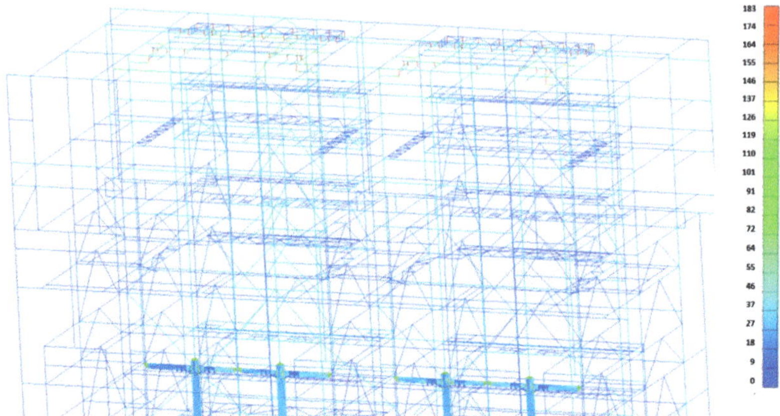

Figure 12. Huber–Mises reduced stress contour lines for the load-bearing structure of the power unit [MPa]—view of the upper grate area.

FEM strength calculations of the load-bearing structure showed among other things that greatest reduced stress equal to 183 MPa could be observed in the area of the posts supporting return lines to the furnace chamber. These values are lower than the yield strength (235 MPa) of the construction material used to build the load-bearing structure of the power units. In the remaining parts of the units' load-bearing structures, the majority of reduced stresses were approximately 50 MPa. The above data prove that the fire in the boiler hall did not cause any damage to the support structure of the power unit. FEM strength calculations were also performed for the container, by building both geometric (Figure 13) and discrete models. The container is supported in four points located in its upper part, as can be seen from the model shown in Figure 13. The colors in the model correspond to metal sheets of different thicknesses [20].

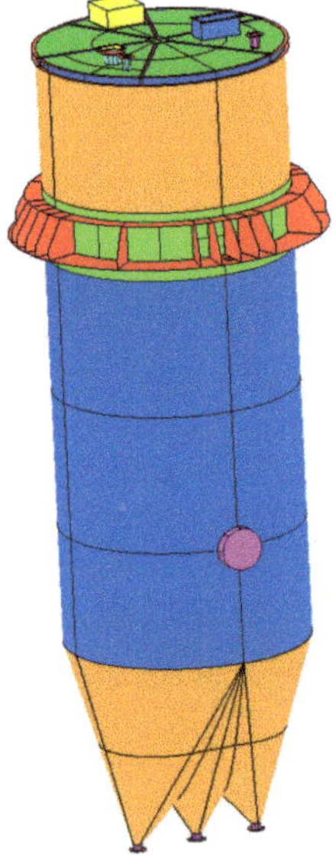

Figure 13. View of the container's complete geometric model.

Digitization was performed using Thin Shell elements whose thickness corresponded to the thickness of metal sheets used to construct the container. The calculations included loads resulting from its normative service load and thermal gradients resulting from local exposure to heat due to the fire wave. The container was made of steel S235, which is why the following strength criteria were assumed:

$$\sigma_{red} < \sigma_{dop} = R_e = 230 \text{ MPa} \tag{3}$$

where:

σ_{red}—reduced stress according to Huber and Mises [MPa];
σ_{dop}—allowable stress assumed in the analysis [MPa];
R_e—yield strength of the container structure's material [MPa].

In order to determine the reasons for the container's damage, a series of strength analyses was performed, with consideration paid to several factors that accompanied the fire and could cause a crack in the area of dumping hoppers and complete damage of the container's upper part. The rising temperature inside the boiler hall, which was the result of the fire, led to uneven heating of the boiler's surface. The thermal gradient of the boiler's casing that occurred in consequence, as well as the rise in air temperature inside the container—which translated into the rise in air pressure—both contributed to the emergence of cracks in the area of weld lines on the fuel hopper and sudden damage of its upper part that followed. Strength analysis was performed for the following air pressure values: 5 kPa, 10 kPa, 15 kPa, ... , 50 kPa (values increased by 5 kPa). Impact strength of approximately 200 MPa was assumed for the observed low quality of weld lines on the fuel hoppers. The strength analysis showed that at air pressure inside the container equal to 50 kPa and at uneven thermal impact that occurred simultaneously, reduced stress in the area of hopper weld lines reached approximately 270 MPa (Figure 14). This contributed to the breaking of the discussed area of the container 8z.

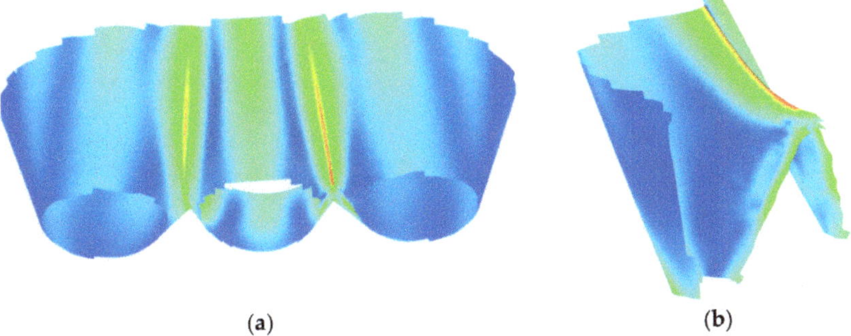

(a)   (b)

Figure 14. Reduced stress contour lines in the lower part of the container, according to Huber and Mises hypothesis [MPa] (**a**) hopper area of the fluidized bed material tank; (**b**) the hopper weld area of the fluidized bed material reservoir where the crack has occurred

At the same time, reduced stress in the container's upper part was observed at approximately 870 MPa (Figure 15). The latter observation, along with the fact that the tests performed prove the ultimate strength of steel S235 to be at the level of 315 MPa, means that this area experienced a sudden discontinuity of the investigated object's casing [20].

The strength analysis also showed clearly that as the air pressure inside the container rose to approximately 50 kPa and the casing heated unevenly, the casing broke in its both upper and lower parts.

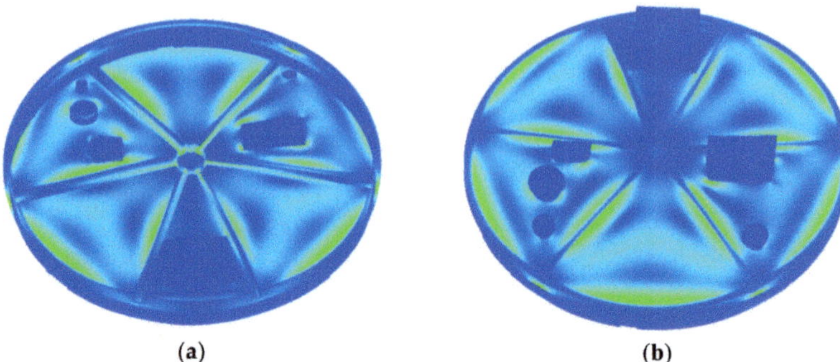

Figure 15. Reduced stress contour lines in the upper part of the container, according to Huber and Mises hypothesis [MPa]: (a) Calculation results of the upper cover of the bed material tank from the outside, (b) Calculation results of the upper cover of the bed material tank from the inside.

7. Summary and Conclusions

Measurement results allowed to estimate the effects of the fire wave that passed through a power station. The works were divided into several stages:

- Investigation of the damage that the passing fire wave caused to the load-bearing structure of the power unit's elements, together with the equipment located in the technological line. As a parallel task, on-site inspections included the power unit's main elements that remained in service for many years.
- Problematic or damaged main elements found during inspection allowed us to estimate the technical condition of the investigated structure and to establish the scope of necessary repair works.
- Inspections of welded joints were performed only in available places, on the unit's main elements, i.e., on the load-bearing structure, second pass chamber, cyclone, ash removal system, air ducts, auxiliary elements, etc. As indicated in the article, much attention was paid to the levels on which the fire wave passed.

The passing fire wave caused greatest damage in front of the unit, at the coal and biomass supply galleries. Coal and biomass ignition in the space between the unit generated soot, which combined with ash and water was used to extinguish the fire and produced compressed mass (Figure 14). The mass had the consistency of concrete and covered electric wires placed in conduits. The wires that were laid in conduits close to each other stopped the flowing mass, which set shortly. This phenomenon was observed only on the 6m level of the power unit [20].

Author Contributions: Conceptualization, M.P.; methodology, A.G.; software, M.P. validation, M.P. formal analysis, A.G., M.P.; investigation, A.G.; resources, A.G.; data curation M.P.; writing—original draft preparation, A.G., M.P.; writing—review and editing, M.P.; visualization A.G., M.P.; supervision A.G.; project administration, A.G.; funding acquisition, A.G. All authors have read and agreed to the published version of the manuscript.

Funding: This research received no external funding.

Data Availability Statement: The data presented in this study are available on request from the corresponding author. The data are not publicly available due to provisions in the contract with the company for which the research was carried out.

Acknowledgments: As part of the work carried out, no support was provided, either from external persons or institutions.

Conflicts of Interest: The authors declare no conflict of interest.

References

1. Schröder-Hinrichs, J.U.; Baldauf, M.; Ghirxi, K.T. Accident investigation reporting deficiencies related to organizational factors in machinery space fires and explosions. *Accid. Anal. Prev.* **2011**, *43*, 1187–1196. [CrossRef] [PubMed]
2. Zhu, Y.; Wang, D.; Shao, Z.; Xu, C.; Zhu, X.; Qi, X.; Liu, F. A statistical analysis of coalmine fires and explosions in China. *Process Saf. Environ. Prot.* **2019**, *121*, 357–366. [CrossRef]
3. Czmochowski, J.; Górski, A.; Iluk, A.; Wyszyński, J. Numeryczna weryfikacja wytężenia przegrzewacza grodziowego kotła energetycznego. *Syst. J. Transdiscip. Syst. Sci.* **2002**, *7*, 105–112.
4. Czmochowski, J.; Górski, A.; Paduchowicz, M.; Rusiński, E. Diagnostic method of measuring hanger rods tension forces in the suspension of the power boilers combustion chamber. *J. Vibroeng.* **2012**, *14*, 129–134.
5. Kowalczyk, M.; Czmochowski, J.; Derlukiewicz, D. Diagnostic Models of the States of Developing Fault for Working Parts of the Excavator. *Solid State Phenom.* **2010**, *165*, 285–289. [CrossRef]
6. Li, L.; Qin, B.; Ma, D.; Zhuo, H.; Liang, H.; Gao, A. Unique spatial methane distribution caused by spontaneous coal combustion in coal mine goafs: An experimental study. *Process Saf. Environ. Prot.* **2018**, *116*, 199–207. [CrossRef]
7. Mi, C.; Li, W.; Xiao, X.; Berto, F. An energy-based approach for fatigue life estimation of welded joints without residual stress through thermal-graphic measurement. *Appl. Sci.* **2019**, *9*, 397. [CrossRef]
8. Coimbra, R.N.; Escapa, C.; Otero, M. Comparative thermogravimetric assessment on the combustion of coal, microalgae biomass and their blend. *Energies* **2019**, *12*, 2962. [CrossRef]
9. Zhou, C.; Gao, F.; Yu, Y.; Zhang, W.; Liu, G. Effects of Gaseous Agents on Trace Element Emission Behavior during Co-combustion of Coal with Biomass. *Energy Fuels* **2020**, *34*, 3843–3849. [CrossRef]
10. Krolczyk, J.B.; Gapiński, B.; Krolczyk, G.M.; Samardžić, I.; Maruda, R.W.; Soucek, K.; Legutko, S.; Nieslony, P.; Javadi, Y.; Stas, L. Topographic inspection as a method of weld joint diagnostic. *Teh. Vjesn.* **2016**, *23*, 301–306. [CrossRef]
11. Landowski, M.; Świerczyńska, A.; Rogalski, G.; Fydrych, D. Autogenous fiber laser welding of 316L austenitic and 2304 lean duplex stainless steels. *Materials* **2020**, *13*, 2930. [CrossRef] [PubMed]
12. Górka, J.; Stano, S. Microstructure and properties of hybrid laser arc welded joints (laser beam-MAG) in thermo-mechanical control processed S700MC steel. *Metals* **2018**, *8*, 132. [CrossRef]
13. Tabatabaeipour, M.; Hettler, J.; Delrue, S.; Van Den Abeele, K. Non-destructive ultrasonic examination of root defects in friction stir welded butt-joints. *NDT E Int.* **2016**, *80*, 23–34. [CrossRef]
14. Vibrations, N.M.; Bergamo, E.; Fasan, M. Efficiency of Coupled Experimental—Numerical Predictive Analyses for Inter-Story Floors Under Non-Isolated Machine-Induced Vibrations. *Actuators* **2020**, *9*, 87.
15. Shittu, S.; Li, G.; Zhao, X.; Akhlaghi, Y.G.; Ma, X.; Yu, M. Comparative study of a concentrated photovoltaic-thermoelectric system with and without flat plate heat pipe. *Energy Convers. Manag.* **2019**, *193*, 1–14. [CrossRef]
16. Ren, K.; Chew, Y.; Zhang, Y.F.; Bi, G.J.; Fuh, J.Y.H. Thermal analyses for optimal scanning pattern evaluation in laser aided additive manufacturing. *J. Mater. Process. Technol.* **2019**, *271*, 178–188. [CrossRef]
17. NX-Ideas. Producer: Siemens PLM Software, Distributor: ATA ENGINEERING INC. Available online: https://www.ata-plmsoftware.com/ (accessed on 22 February 2021).
18. Doronin, S.V.; Reizmunt, E.M. Multi-Model Interval Estimators of the Strength State of Structurally Complex Load-Bearing Units of Production Equipment. *Chem. Pet. Eng.* **2018**, *53*, 604–609. [CrossRef]
19. Ortlepp, R.; Ortlepp, S. Textile reinforced concrete for strengthening of RC columns: A contribution to resource conservation through the preservation of structures. *Constr. Build. Mater.* **2017**, *132*, 150–160. [CrossRef]
20. Liu, J.; Xu, T.; Guo, Y.; Wang, X.; Frank Chen, Y. Behavior of circular CFRP-steel composite tubed high-strength concrete columns under axial compression. *Compos. Struct.* **2019**, *211*, 596–609. [CrossRef]

Article

Investigation of Waste Biogas Flame Stability Under Oxygen or Hydrogen-Enriched Conditions

Nerijus Striūgas, Rolandas Paulauskas *, Raminta Skvorčinskienė and Aurimas Lisauskas

Laboratory of Combustion Processes, Lithuanian Energy Institute, Breslaujos str. 3, LT-44403 Kaunas, Lithuania; Nerijus.Striugas@lei.lt (N.S.); Raminta.Skvorcinskiene@lei.lt (R.S.); Aurimas.Lisauskas@lei.lt (A.L.)
* Correspondence: Rolandas.Paulauskas@lei.lt; Tel.: +370-37-401830

Received: 18 August 2020; Accepted: 9 September 2020; Published: 11 September 2020

Abstract: Increasing production rates of the biomethane lead to increased generation of waste biogases. These gases should be utilized on-site to avoid pollutant emissions to the atmosphere. This study presents a flexible swirl burner (~100 kW) with an adiabatic chamber capable of burning unstable composition waste biogases. The main combustion parameters and chemiluminescence emission spectrums were examined by burning waste biogases containing from 5 to 30 vol% of CH_4 in CO_2 under air, O_2-enriched atmosphere, or with the addition of hydrogen. The tested burner ensured stable combustion of waste biogases with CH_4 content not less than 20 vol%. The addition of up to 5 vol% of H_2 expanded flammability limits, and stable combustion of the mixtures with CH_4 content of 15 vol% was achieved. The burner flexibility to work under O_2-enriched air conditions showed more promising results, and the flammability limit was expanded up to 5 vol% of CH_4 in CO_2. However, the combustion under O_2-enriched conditions led to increased NOx emissions (up to 1100 ppm). Besides, based on chemiluminescence emission spectrums, a linear correlation between the spectral intensity ratio of OH* and CH* (I_{OH*}/I_{CH*}) and CH_4 content in CO_2 was presented, which predicts blow-off limits burning waste biogases under different H_2 or O_2 enrichments.

Keywords: swirl burner; waste biogas; hydrogen; oxygen; combustion; flame stability; blow-off limit

1. Introduction

Due to concerns regarding global climate change, there has been a search for new ways to improve energy efficiency, reduce pollutant emission, and ensure sustainable management of resources. For these reasons, various alternative fuels and by-products (gas or waste fuel) with low calorific value are gaining more attention as an alternative energy source for energy production [1–3]. Basically, such fuels are gained in different fields with different compositions like steel production, refineries, landfills, biogas plants, gasification plants, and other industry sources. For example, a blast furnace gas or syngas consists of different concentrations of CO, H_2, CO_2, and/or N_2, while biogas is mainly composed of CO_2 and CH_4. Considering that the biogas and biomethane production is increasing over the years, rates of the waste biogas from biomethane production are also growing. These gases mainly consist of from few to 25 vol% of methane and should be utilized on-site to prevent pollutant emissions to the atmosphere. However, these gases are not well suitable to combust in existing natural gas-fired systems as the low concentrations of CH_4, and the high amount of CO_2 could result in an unstable flame, blow-off, and a release of unburned gases into the environment. Mainly it caused by the high amount of CO_2, which affects the flame temperature, thereby reducing the stability of the flame.

To avoid these problems and ensure complete and clean combustion of waste biogases, flame stability must be improved, and flammability limits expanded. It could be achieved using new concepts/modifications of burners [4–7]. For example, a new burner design based on modified geometry was proposed and tested for low calorific gas combustion by [8]. It was found that this burner ensures stable combustion of natural gas and syngas blends, but CO emissions exceed 200 mg/m^3 if thermal shares of syngas are below 70%. Another concept burner based on continuous air staging for low calorific gases was developed by a research group from Gaswärme-Institute. V. Essen (GWI) [9]. The burner (up to 200 kW) was tested on syngas, landfill gas, and mine gas. The obtained results showed that it is possible to achieve a stable combustion process with low NOx and CO emissions, but there is a need to preheat these gases up to 400 °C before supplying to the combustion zone. A similar study was performed by Mortberg et al. [10]. The authors preheated air up to 900 °C to enhance the low calorific value (LCV) gas combustion stability. During combustion tests, a cross-flow jet of LCV gas mixture (11.9 vol% CH_4 in N_2) was introduced in the combustion airflow at oxygen-deficient conditions. The obtained results showed that this configuration using preheated combustion air is suitable to combust LCV gases, but results in the prolonged ignition delay, higher turbulence levels, and higher vorticity. Another work [11] showed that surface-stabilized combustion (SSC) technology burner is able to burn the LCV biogas with 20–65 vol% CO_2 in CH_4 or hydrogen-enriched CH_4 (15–100 vol% H_2 in CH_4) with low pollutant emissions. However, the burner was more suitable for combustion of high calorific value gases. In the case of biogas with 65 vol% CO_2 in CH_4, the flames were lifted, changing φ from 0.98 to 0.8, while further increase in φ led to blow-off. Another research [12] also focused on the combustion of LCV gases in burners with SiC and Al_2O_3 porous structures. The authors were capable of combusting landfill gas with a methane content of 26 vol%. Though, at higher power (10 kW), the flammability limit decreased by 4 vol% of CH_4 in CO_2 due to increased flow speed. The authors also noticed that preheating the gas mixture, and the flammability limit could be extended by 2 vol%. Al-Attab et al. [13] investigated the combustion of producer gas from biomass gasification in a two-layer porous burner. Results showed that the burner is able to combust the producer gas with a lower calorific value of 5 MJ/m^3 changing equivalence ratios φ in the range of 0.33–0.71, but NOx emissions were in the range from 230 to 270 ppm. In order to increase the blow-off limit of LCV gases, Song et al. [14] investigated an improved preheating method of the gas mixture, based on an annular heat recirculation. The gas flow was preheated by a high-temperature wall near the gas inlet, which temperature increase, due to improved axial heat conduction and radiation heat transfer of porous media. It ensured stable combustion of the LCV gas of 1.4 MJ/m^3. However, according to [15], higher volumetric capacity is needed to achieve higher power comparing to existing conventional burning systems. Besides, the use of porous media burners for the combustion of waste biogases with the varying composition is a bit complicated as a material for the porous media should be designed considering gases needed to burn; otherwise, a periodic replacement of the porous media could be needed.

Another option to improve flame stability and flammability is to use an addition of hydrogen/syngas or supply of oxygen-enriched air [16–22]. Chiu et al. [23] investigated the effect of H_2/CO on a premixed methane flame. The research was performed in an impinging burner by changing an H_2/CO concentration from 20/80 to 80/20 vol% and a methane concentration from 10 to 20 vol%. It was determined that the stable flame with H_2/CO of 20/80 vol% is achieved at reach combustion conditions ($\varphi \geq 1.8$) though the flame stability and flammability increased with increasing the hydrogen concentration in simulated syngas. It was also noted that with the increasing amount of H_2 in CO increases the flame temperature. A similar work focused on the hydrogen effect on the LCV gas combustion was performed by [24]. The authors performed numerical and experimental burner tests at a thermal power of 10 kW using syngas (~4.7 MJ/m^3) and blast-furnace gas (~3.7 MJ/m^3), including the addition of H_2 (26 and 52 vol%). It was determined that the burner properly burns the syngas, but the blast-furnace gas requires an additional supply of H_2 to avoid blow-off. Though, it was observed that the addition of H_2 negatively affects NOx emissions. To extend LCV gas combustion flammability and improve flame stability, this burner was also tested under oxy-fuel combustion conditions [25]. Results showed that NOx emissions decrease; however, to achieve a more stable flame and avoid overheating of the burner wall, a modification for the oxidizer distribution is needed. More detailed investigation of the oxygen enrichment was performed by Ylmaz et al. [26]. The authors investigated the biogas flame stability and formation of emissions under different oxygen enrichment conditions in a pilot-scale model burner. It was pointed out that an increase of O_2 to 24% in the oxidizer leads to improved flame stability though CO emissions were equal to 372 ppm. A higher enrichment level (28 vol% of O_2) caused decreased flame stability, but lower CO emissions, till 10 ppm. During the tests, NOx emissions increased with increased O_2 concentrations in the oxidizer. Ba et al. [20] tried to improve flame stability by coupling oxy-fuel combustion and fuel/air preheating. Firstly, this configuration was experimentally tested using a lab-scale tri-coaxial burner (25 kW), and later, the burner scaled-up to 180 kW was tested on a semi-industrial facility. In both cases, a stable flame of blast furnace gas was achieved with very low levels of CO and NOx emissions. However, this configuration was not tested on waste biogases, and the effect on flame stability is unknown.

Literature review reveals that there are many works proposing techniques to ensure stable combustion of syngas, furnace blast gases, natural gas blends with LCV gases and etc., but studies related to the combustion characteristics of waste biogas are still lacking. Taking into account that biogas production rates are increasing and to prevent waste biogas emissions to the atmosphere, these gases should be utilized safely on-site. The previous work [27] showed that the addition of up to 20 vol% hydrogen or oxygen-enriched air ensures stable combustion of waste biogases ($CH_4 > 30\%$ in CO_2) with low emissions in a flat flame burner. Based on these findings, a flexible swirl burner (~100 kW) with an adiabatic chamber capable of burning unstable composition waste biogases was designed and developed. The developed combustion system provides a solution to recover primary energy and reduce pollutant emissions from biomethane production obtained waste biogases. This study presents the flexible combustion system in detail and performed tests in real conditions. During the experimental tests, flame stability and combustion limits were explored to ensure a wide range operation of the burner with different waste biogas containing from 5 to 30 vol% of CH_4 in CO_2 under air, oxygen-enriched atmosphere, or with hydrogen addition. For a better understanding of waste gas flame behavior under different conditions and to identify a burner performance via flame transition modes and blow-off limits, chemiluminescence emission spectrums (OH* and CH*) from flames of different waste biogas mixtures were obtained. During all tests, an online flue gas analysis was also performed to determine variations of NOx and CO emissions and compliance with emission standards.

2. Materials and Methods

2.1. Experimental Setup

The experiments were carried out in a 200 kW$_{th}$ semi-industrial scale test facility (Figure 1). In order to simulate waste biogases (LCV) with a low content of methane (CH_4) in carbon dioxide (CO_2) and its combustion with oxygen (O_2) or hydrogen (H_2) addition, a gas supply system was constructed (1–5). CO_2 was acquired in a liquid state (1) by a local supplier (JSC Gaschema), directed to a vaporizer (2) and supplied via the control unit to a gas-mixing system for preparation of LCV gases with pre-assigned compositions. Methane in the form of natural gas (NG) with the composition of 97.5 vol% CH_4, 1.2 vol% C_2H_6, 0.27 vol% C_3H_8 and C_4H_{10}, and 0.08 vol% CO_2 (NG composition is taken from the gas supplier's datasheet) was delivered from a distributed NG pipeline and supplied through a pressure regulating and metering station (3) to the gas-mixing system. For hydrogen or oxygen-enriched combustion, these gases were supplied from gas cylinders (4,5) to a novel swirl LCV gas burner (7). The required gas flow was ensured by a mass flow controller. The combustion air was supplied using an air blower (6).

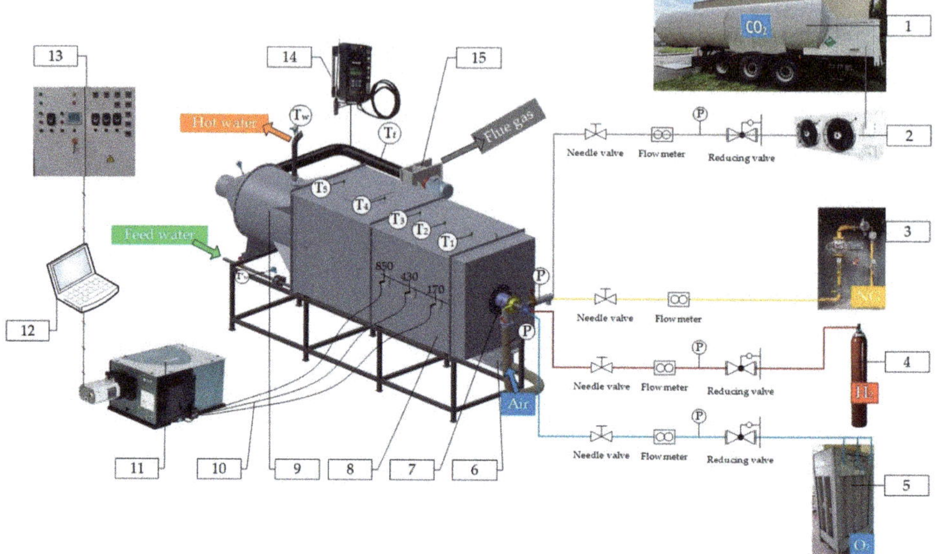

Figure 1. Scheme of the experimental burner testing setup: 1–A liquid CO_2 tank; 2–a CO_2 vaporizer; 3–a natural gas pressure regulating and metering station; 4–a hydrogen cylinder; 5–an O_2 cylinder bundle with a manifold; 6–an air blower; 7–a novel low calorific value (LCV) gas burner; 8–an adiabatic combustion chamber; 9–a boiler; 10–optical fibers with collimating lens; 11–a spectrometer with an ICCD camera; 12–a data collecting system; 13-an operation panel; 14–a flue gas analysis system; 15–a flue gas blower.

The combustion of the prepared LCV gas mixture was performed in an adiabatic chamber (8) with internal dimensions of $600 \times 600 \times 2500$ mm. The temperatures in the chamber were measured with four K type thermocouples (T_1–T_4). The thermocouples were positioned at the 530, 900, 1690, and 2410 mm from the inner front wall, respectively. For spectral analysis of selected excited radicals from the flame, a spectrometer Andor Shamrock SR-303i coupled with an ICCD (Intensified Charge Coupled Device) camera Andor iStar DH734 (11) was used. Three optical fibers with collimating lenses (10) were installed to transmit the light to the spectrometer. The optical fibers were oriented horizontally to provide a radical distribution along with the flame and fitted at the distances of 170, 410, and 830 mm from the front wall. The combustion products are cooled down to appropriate temperature, which is measured with a K type thermocouple, by passing the boiler (9) mounted at the exit of the combustion chamber. A flow of cooling water is controlled, and water temperature is monitored before and after the boiler. The cooled flue gases are analyzed using a portable flue gas analyzer Testo 350XL (14) via a measurement point installed between the stack (15) and the boiler (9). The entire combustion process is automated and controlled from a control desk (13) installed outside the site. All operating signals from measuring equipment or sensors were collected and analyzed further.

2.2. Waste Biogas Burner

A novel swirl burner was developed for the waste biogas (LCV) combustion (Figure 2). The burner is designed to operate in a flexible mode for a variety of waste biogas compositions, including the possibility of hydrogen or oxygen addition. The burner has multiple inlets for different gases and operational regimes. The ignition of the main fuel (LCV gas) is executed with a pilot natural gas flame. After the ignition of the main fuel, a supply of pilot gas is stopped. The LCV gas is supplied via an annular channel installed in the center of the burner. Air was used as the main oxidizer, which enters the burner through a peripheral annulus. When a calorific value of waste biogas becomes very low, a stable operation is ensured by the oxygen or hydrogen addition. The supply of these gases is designed in the center of the burner.

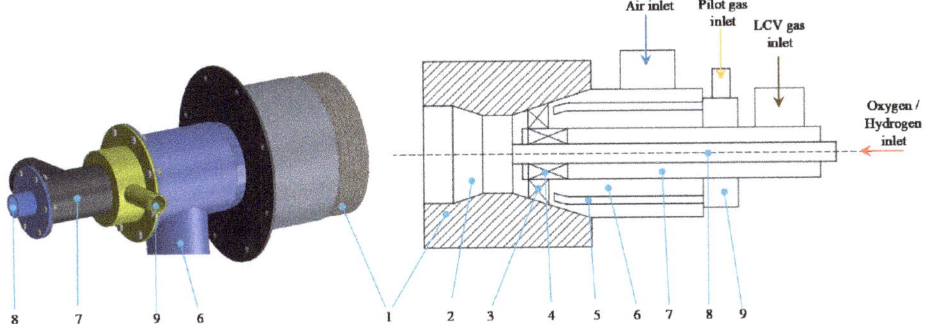

Figure 2. Schematic view of the experimental LCV gas burner:1–A burner embrasure; 2–a burner quarl; 3–an axial airflow swirler; 4–an axial LCV gas swirler; 5–a pilot gas lance; 6–an air annulus; 7–an LCV gas annulus; 8–an oxygen or hydrogen orifice; 9–a pilot gas annulus.

The flame stabilization is achieved by swirling both, the combustion air and LCV gas flow with an axial vane swirl angle of 30°. Both swirlers has a swirl number of $S_w = 0.5$, which was defined by Equation (1) [28]:

$$S_w = \frac{2}{3} \tan(\alpha_{sw}) \left(\frac{1 - R^3}{1 - R^2} \right) \tag{1}$$

where α_{sw} is the axial vane swirl angle, R is the ratio of the center-body diameter and outer inlet tube diameter.

2.3. Experimental Procedure

The stability limits of waste biogas (LCV) combustion were determined by running the combustor at a fixed heat load of 80 kW_{th}. The heat load was calculated using the lower heating values (LHV) and measuring flows of combustible gases, namely, CH_4, and H_2 when it was used. Thus, the main variables were CO_2 content in LCV gas, an amount of the oxidizer, and the fuel to air ratio (φ).

Three sets of experiments were performed to determine the limits of stable waste biogas (LCV) combustion. The first trial was associated with a determination of the lowest possible, stable combustion regime by decreasing LHV of LCV gas, which was changed by increasing the volumetric fraction of CO_2. The air was used as the oxidizer. At each LCV gas composition, the burner performance was tested for different fuel to air ratios, which varied in the range from 0.7 to 1.0. The fuel to air ratio was determined by measuring the O_2 concentration in flue gases.

The second round of experiments was conducted with the H_2 addition to extend LCV gas combustion limits. In a previous work [29], the possibility to produce H_2 on-site from LCV gas was explored. The idea was based on the power of syngas (P2SG) technology. Therein to produce H_2 rich syngas, the thermal plasma was used to run a dry reforming of gases containing a low concentration of CH_4 in CO_2. Finally, it was concluded that from an economic point of view, this technology is cost competitive when a produced H_2 content for 1 m^3 LCV gas combustion will constitute up to 5 vol%. To simulate this situation in the present work, the hydrogen addition of 2.5 and 5 vol% was investigated. As in the previous case, the air was used as the oxidizer with ϕ ranging from 0.7 to 1.0.

During the third set of experiments, the stability limits were determined using O_2-enriched air instead of the H_2 addition. The amount of oxygen was gradually increased until the flame becomes stable to avoid possible problems described in some works [21,26] like increased CO emissions, high flame temperature, or even unstable combustion. When the flame became stable, the burner performance for different LCV gases against the various fuel to air ratios from 0.7 to 1.0 was tested. The expression of the volumetric oxygen content in this work corresponds to the O_2 concentration in air, which was defined as follows:

$$O_2 \text{ in air} = \frac{V_{air}0.21 + V_{O_2}}{V_{air} + V_{O_2}} \tag{2}$$

where V_{air} and V_{O_2} are the volume flow rates of air and O_2, respectively. The upper value of O_2 in the air was pre-selected to not exceed 30 vol%.

During all experimental runs, emissions of CO and NOx in flue gases were measured continuously. CO is considered as the main indicator for incomplete combustion, thus despite the existence of flame stability, it demonstrates the inappropriate burner operation regime. NOx is one of the main regulated gaseous pollutants, which should be controlled. It is well known, that the main source of NOx formation is attributed to the thermal pathway. Thus, increasing oxygen content in the oxidizer during LCV gas combustion could lead to noticeable higher NOx emissions [3]. In order to know the level of this pollutant during combustion and compliance with emission standards (EU directive 2015/2193 [30]), it was measured and analyzed regarding the LCV combustion mode.

In addition, the chemiluminescence based optical diagnostic tool was used to identify the possible flame transition mode. The main chemiluminescent species of interest were OH* and CH* with a wavelength of 308.9 ± 2.0 nm and 431.4 ± 2.0 nm, respectively. According to [31–34], CH* formation is related to high temperature, and the main formation pathways are:

$$C_2 + OH \rightarrow CH^* + CO, \tag{3}$$

$$C_2H + O \rightarrow CO + CH^*, \tag{4}$$

$$C_2H + O_2 \rightarrow CO_2 + CH^*, \tag{5}$$

The formation and excitation of the OH* radical are also attributed to the thermal excitation, and the main formation reactions are [31–34]:

$$H + O + M \rightarrow OH^* + M, \tag{6}$$

$$CH + O_2 \rightarrow OH^* + CO, \tag{7}$$

The optical data were collected from three-measurement points using a spectrometry system (Figure 1). The light coming from the flame is dispersed, and entire spectra is collected on the split area of the ICCD camera sensor. The camera exposure duration was set to 3 s, a side input slit to 100 µm, and grating to 300 L/mm. Each spectrum consisted of 10 acquisitions.

3. Results and Discussion

3.1. Stability Maps of Waste Biogas (LCV) Combustion

This section presents combustion stability maps for developed LCV gas burner under different tested combustion modes. Firstly, LCV gases with different concentrations of CH_4 and CO_2 were burned in the air over a wide range of fuel to air ratios to determine the limit of stable combustion under ordinary conditions. As can be seen from Figure 3, the stable combustion of the waste biogas (LCV) with $CH_4 \geq 25$ vol% in CO_2 can be reached in the air under all tested range of ϕ. Compared to previous results [27], the combustion of waste biogas in a flat flame burner under the air atmosphere was only able with a mixture of 30 vol% CH_4 in CO_2. In that case, the developed burner shows wider flammability ranges under air conditions. A reduced CH_4 content to 20 vol% in the waste biogas (LCV) leads to flame stability problems, and blow-off occurs at a richer combustion mode,

$\phi \geq 0.87$. According to obtained results (Figure 3), the stable combustion of LCV gases with lower CH_4 content (>20 vol%) is only possible with H_2 or O_2 addition. The addition of H_2 to the biogas flame was studied in detail by others [16,35,36]. Zhen et al. [35] noted that the H_2 addition of 5 vol% improves biogas flame stability, and CO emissions decrease, but further addition increase to 10 vol% of H_2 is insignificant on flame stability. Similar findings were also reported by Leung and Wierzba [17]. The authors investigated the effect of 10 vol%, 20 vol%, and 30 vol% H_2 addition on the biogas flame stability and determined that the most significant results are achieved by supplying a small amount of H_2.

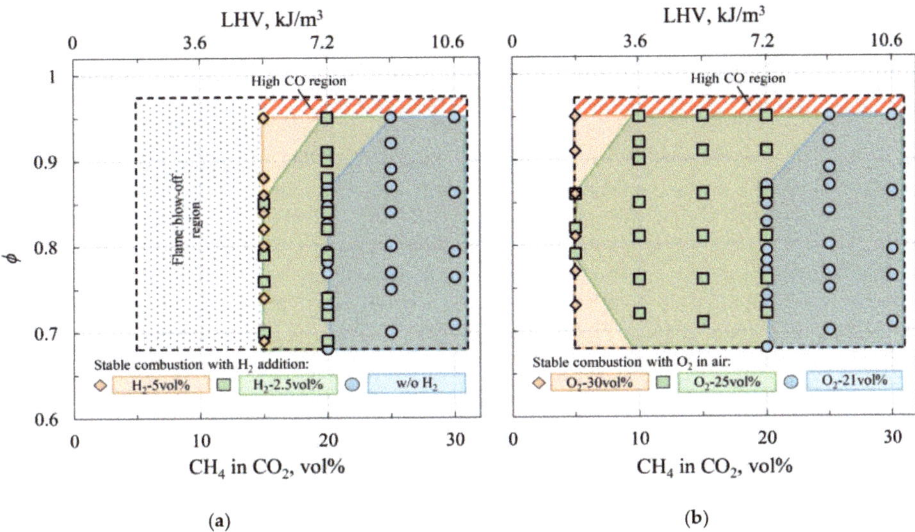

Figure 3. Stability maps of waste biogas (LCV) combustion with different content of CH_4 in CO_2 for a range of (a) H_2 addition and (b) O_2 air enrichment.

Even though the concentration of CH_4 in biogas is higher and varies from 40% to 60% compared to the studied range of CH_4 concentrations (5–30 vol%), the provided findings are in good agreement with the obtained results. The stable combustion of LCV gas containing 15 vol% of CH_4 is obtained within the range of ϕ from 0.7 to 0.85 by adding H_2 of 2.5vol%, while an increased addition of H_2 to 5 vol% has an insignificant effect, and the flammability limit is extended only to a richer region, up to ϕ of 0.95. However, further decrease of CH_4 content in LCV gases results in flame blow-off even in both cases of H_2 addition (Figure 3a). In order to extend the flammability limit and ensure stable combustion of LCV gases with very low CH_4 content (5–15 vol%), O_2-enriched air was introduced. Considering that a high oxygen enrichment level could lead to the unstable flame and increased CO emissions [26], two cases with 25 vol% and 30 vol% of O_2 were tested (Figure 3b). As was expected, using O_2-enriched air with O_2 content of 25 vol%, the stable combustion of waste biogases with CH_4 content from 30 to 5 vol% was ensured. However, in this case, the stable combustion of LCV gas containing 5 vol% of CH_4 is achieved only in the narrow range of ϕ, from 0.79 to 0.86. Meanwhile, at a higher enrichment level (30 vol% of O_2), the stable combustion of LCV gases is acquired in all tested points (Figure 3b). Similar results were also acquired numerically studying methane combustion under CO_2/O_2 atmosphere, and researchers pointed out that the most effective and stable combustion is achieved at O_2 concentrations of 28–32 vol% [37]. Though another work [21] showed a bit higher level of O_2 (31–35 vol%) is needed to achieve a stable flame. Considering the obtained results, it could be assumed that the designed burned with the adiabatic chamber ensure stable combustion of waste biogases (5–30 vol% of CH_4 in CO_2) at lower concentrations of O_2 (25–30 vol%), due to a special design of the burner allowing the faster mixing of fuel oxidizer, faster reaction kinetic and minimal heat loss via the adiabatic combustion chamber walls. However, the presented design of the LCV gas burner shows an opposite relation in non-premixed combustion. The flame extinguish is more likely at higher fuel to air ratios than at lower ones, and this was found for all studied cases (Figure 3). Typically, the flame is more stable, or a flash-back phenomenon is more likely at lower fuel to air ratios, even in premixed flames [25]. The reasons for an instability arose might be related to ongoing several complex processes: The aerodynamic of the flame shape, which allows a proper internal hot gas recirculation zone, and an amount of heat provided from both recirculating gases and re-radiation from furnace walls [37].

In order to study the impact of those two parameters, the flame core temperature and averaged axial flow speed were analyzed. Figure 4 represents data obtained by T_1 thermocouple and shows temperature ranges during the combustion of waste biogases (LCV) with different content of CH_4 in CO_2 for all three tested cases: Under air only combustion, under H_2 addition, and under O_2-enriched air. The upper and bottom limits stand for the maximum (ϕ_{max}) and minimum (ϕ_{min}) fuel to air ratio ϕ at which stable combustion still occurs. As can be seen from the temperature map (Figure 4a,b), the temperature in the core of the flame also rises with an increase of ϕ. While in opposite the temperature decreases with a decrease of CH_4 content in CO_2. The latter trends are well known and logical. However, the experiments showed that the temperature increase not always plays a key role and is sufficient to maintain a stable flame. Meanwhile, considering the axial velocity ranges (Figure 5) can be assumed that a change in velocities is contrary to the temperature and a decrease at higher ϕ ratios occurs. Based on this can be stated that a reduction in flow velocity affects an inner recirculation zone, and thus, prevents passing a sufficient amount of high temperature gas, which in turn negatively affects the combustion stability. This is well proven comparing the cases with H_2 addition and O_2 enrichment at 15 vol% of CH_4 in CO_2 as the flame temperature was higher almost by 100 °C under the oxygen-enriched conditions than in the case of H_2 addition (Figure 4a,b). Though, the flow velocities were in the same order of magnitude. Taking into account these findings, it can be concluded that a certain amount of heat is critical and should be ensured and maintained to ignite the incoming cold air-fuel mixture.

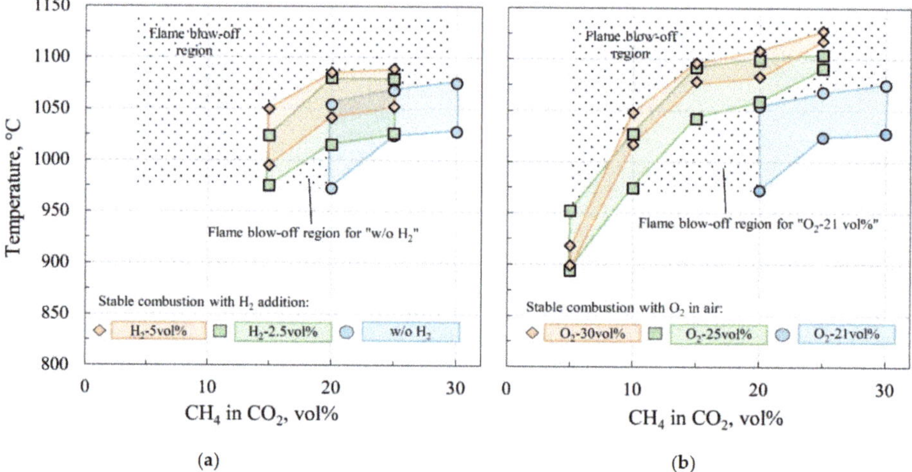

Figure 4. Temperature ranges during waste biogas (LCV) combustion with different content of CH_4 in CO_2 for a range of (**a**) H_2 addition and (**b**) O_2 air enrichment. The upper and bottom limit stands for a ϕ_{max} and ϕ_{min}, respectively.

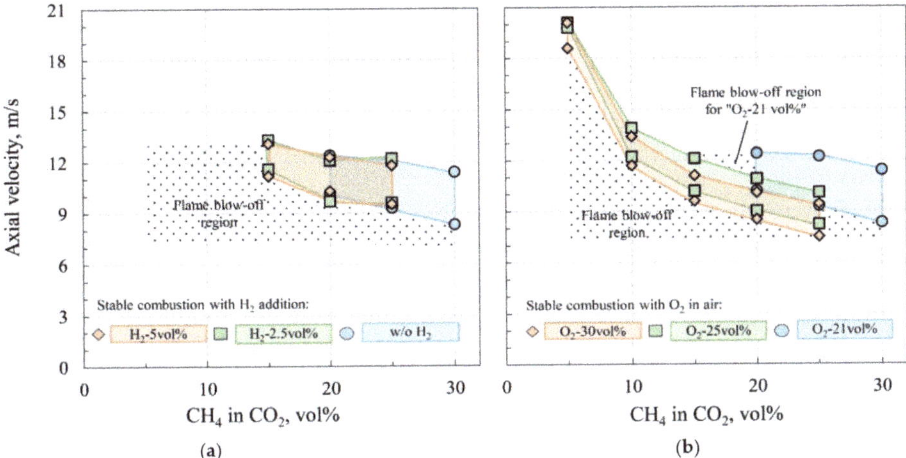

Figure 5. Axial velocity ranges during waste biogas (LCV) combustion with different content of CH_4 in CO_2 for a range of (**a**) H_2 addition and (**b**) O_2 air enrichment. The upper and bottom limit stands for a ϕ_{max} and ϕ_{min}, respectively.

3.2. Changes in Flame Chemiluminescence

For a better understanding of waste biogas (LCV) flame behavior under H_2 or O_2 addition and identify a possible performance for flame transition modes, chemiluminescence emission spectrums from flames of different LCV gas mixtures were obtained. The obtained OH* emission intensities distributed per the chamber length at $\phi = 0.85$ are presented in Figure 6. The highest OH* intensities were observed at the front of the flame (17 cm from the burner) where the main reactions take place. Comparing cases with and without H_2 enrichment, it was observed that the H_2 addition of 2.5 vol% has a negligible effect on the LCV flame enhancement, especially for the mixture with 25 vol% CH_4 in CO_2 as the combustion chamber temperatures are near identical to ones without the H_2 addition (Figure 4a). Similar results were also determined analyzing the obtained OH* emission intensities of hydrogen-enriched LCV flames (Figure 6a). Though, the increase of hydrogen addition to 5 vol% led to higher OH* emission intensities by ~1.6 times and a higher chamber temperature at point T_1 by ~30 °C burning LCV gas mixtures with CH_4 content of 25 vol% and 20 vol%. According to previous work [27], the increased OH* intensities and chamber temperatures could be attributed to an improved flame stability. However, the waste biogas (LCV) flammability limit using the H_2 enrichment was achieved burning mixtures with CH_4 content of 15 vol%, and a further decrease in CH_4 content led to the flame blow-off. In the case of the mixture with 15 vol% CH_4, lowermost OH* emission intensities, and chamber temperatures were determined using both H_2 additions (2.5 and 5 vol%). This could be related to a high level of the diluent (CO_2), which in turn lowers the flame temperature, and thus, OH* emission. The latter findings are in agreement with those found in the work of Guiberi et al. [38], where pointed out that OH* intensity decrease of CO_2-diluted flame is more intense with the increasing diluent concentration.

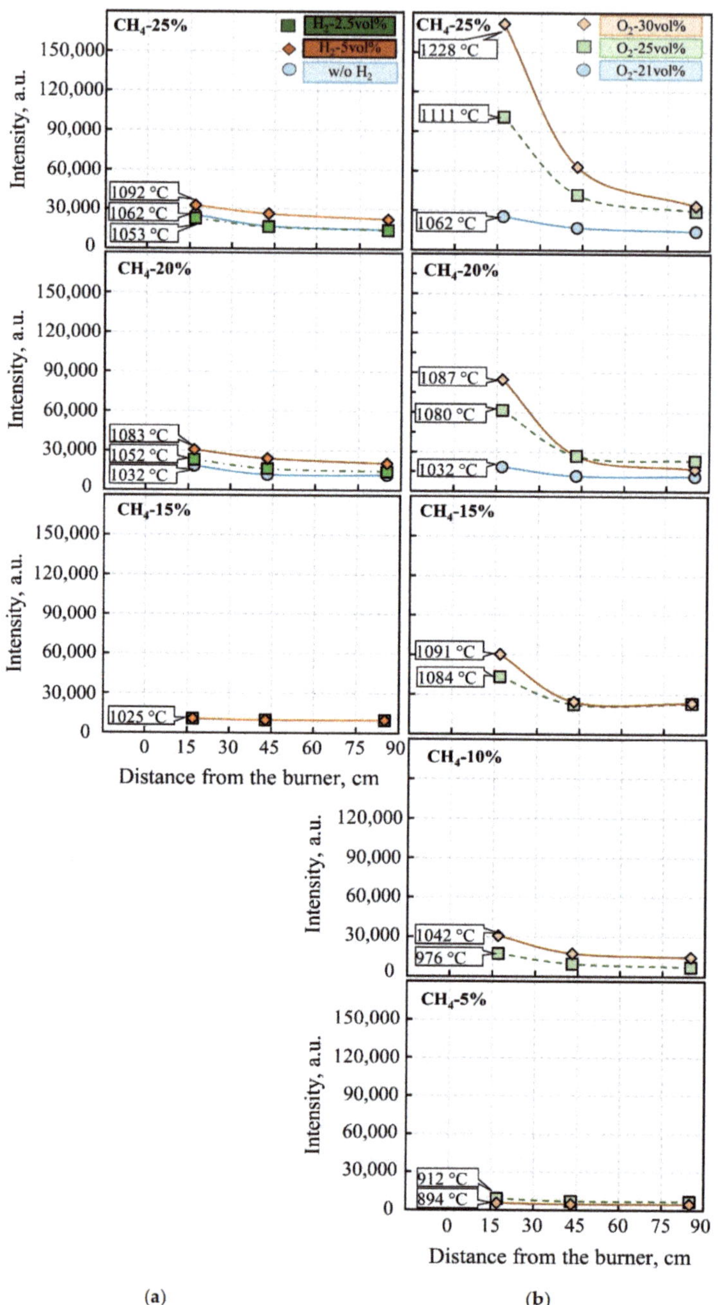

Figure 6. OH* spectral intensity distribution per distance at $\phi = 0.85$ under (**a**) hydrogen-enrichment and (**b**) oxygen-enriched air.

A bit different results were obtained using oxygen-enriched air (Figure 6b). As an example, an increased O_2 content by 4 vol% in the air affected the mixture with 25 vol% of CH_4 combustion process significantly, and the OH* emission intensity at the flame front was almost four times higher compared to the intensity obtained under ordinary combustion in air. Besides, higher chamber temperatures were also observed (Figure 6b). Further increase in O_2 content from 25 to 30 vol% also led to increased OH* emission intensities and the chamber temperatures. A similar trend of OH* emission intensity increase, due to a higher O_2 content in the oxidizer was also determined in previous works [27,39]. According to He et al. [18], the OH* formation reaction (7) dominates under oxygen-enriched conditions as the intensity of reaction (7) is enhanced when the concentration of O_2 in the oxidizer increases and vice versa. However, with increasing concentration of diluent or decreasing concentration of methane, the OH* emissions weaken gradually, resulting in decreased spectrum intensity. This tendency was determined by burning LCV gases with lower content of CH_4 (20–5 vol%). Even though the supply of oxygen-enriched air expanded LCV gas flammability and was able to burn the mixture with CH_4 content of 10 and 5 vol%.

At the lowest CH_4 concentration (5 vol%) under oxygen-enriched air (25 and 30 vol% of O_2), OH* emission intensities were lower than that in the case of the mixture with 20 vol% CH_4 burned under air combustion mode. Though the chamber temperature at point T_1 was about 900 °C, the combustion process was stable using both oxygen-enrichments, possibly due to enhanced fuel oxidation. According to obtained results under oxygen-enriched conditions, the OH* intensity decrease coincides with the decrease of chamber temperatures, and the OH* emission intensities are related to the flame temperature. In overall, this tendency is in good agreement with spectral intensity ratios of I_{OH*}/I_{CH*} changing CH_4 content in CO_2 (Figure 7) even though the ratio of I_{OH*}/I_{CH*} is mostly used to determine the global equivalence ratio [40,41]. Increasing the CO_2 dilution level leads to decreased flame temperature as the LHV decreases, and the ratio of I_{OH*}/I_{CH*} also decreases. The supply of O_2-enriched air results in rapidly increased ratios of I_{OH*}/I_{CH*}, but it also depends on the O_2 level in the air and CO_2 levels in LCV gases (Figure 7). Considering the stability of waste biogas (LCV) combustion, the ratios of I_{OH*}/I_{CH*} determined burning LCV gases under air combustion conditions could be assumed as indicators for a threshold of the stable combustion as a further decrease in the ratios of I_{OH*}/I_{CH*} lead to blow-off. Under oxygen-enriched conditions, the lowest ratio of I_{OH*}/I_{CH*} is achieved by burning LCV gas with the lowest CH_4 content (5 vol%), and a further decrease in CH_4 content or O_2 enrichment level also leads to the flame extinguish and blow-off. Taking into account these observations, a dashed line in black was introduced in Figure 7 to represent the stability threshold of LCV gas combustion. The ratios of I_{OH*}/I_{CH*} on or above the dashed line show the stable combustion, and the combustion stability the increases with increase of CH_4 content in CO_2 and O_2 level in the air. Meanwhile, the flame blow-off occurs below the stability line. Besides, the proposed indication of the stable combustion is in close agreement with the determined stability maps (Figure 4b). However, this tendency is not fully valid for the cases with H_2 addition. This could be related to low content of H_2 addition, as it might not be sufficient to represent the flame enhancement via flame chemiluminescence. The previous research showed similar results with the same level of H_2 addition, and the significant combustion improvement and higher intensities of OH* and CH* were achieved only at higher levels of H_2 addition [27].

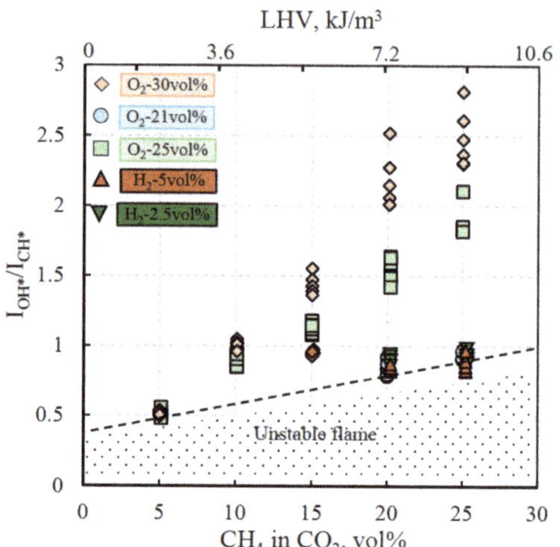

Figure 7. The tendency of the spectral intensity ratio of I_{OH*}/I_{CH*} with the increase of CH_4 content in CO_2.

3.3. Addictive Influence on NOx and CO Emission

During the waste biogas (LCV) combustion tests, an online flue gas analysis was performed to determine the behavior of NOx and CO emissions under all tested cases. The obtained concentrations of NOx and CO are presented graphically in Figure 8. The burner configuration ensured low NOx (up to 25 ppm) and zero CO emissions burning LCV gases with CH_4 content of 25 and 20 vol% with air in the range of ϕ from 0.7 to 0.9. However, as noted before, the combustion at higher ϕ values (>0.9) becomes unstable, and CO is formed, in which concentration in the exhaust gases rises up to 45 ppm. This trend was determined in all tested conditions. At ϕ values below 0.9, the stable combustion is achieved, and variations in CO and NOx emissions mainly depend on the CH_4 content in LCV gases and the level of O_2 or H_2 addition. In the case of H_2 addition (Figure 8a), the NOx concentrations increased with increasing addition of H_2 compared to emissions obtained at air combustion mode. For example, in the case of the mixture with 20 vol% CH_4, the NOx concentrations increased by 2–5 ppm and by 5–7 ppm using the H_2 addition of 2.5 vol% and 5 vol%, respectively. According to Figure 4a, it could be considered that NOx concentrations increase due to increased flame temperature, which in turn is affected by enhanced flame speed, and thermal NOx formation intensifies with higher levels of H_2 addition. Meanwhile, the CO concentrations stand at zero, due to enhanced combustion. It is important to point out, that the lowest NOx (up to 14 ppm) and zero CO concentrations using the H_2 addition are obtained burning the LCV gases with CH_4 content of 15 vol%. In comparison, this is not able to combust under normal conditions (Figure 8a). Besides, the obtained emissions of NOx during the waste biogas combustion with hydrogen addition does not exceed the emission limit value (97 ppm) according to EU directive 2015/2193 [30]. Considering the oxygen-enriched conditions, the stable combustion of the 15 vol% CH_4 in CO_2 is also achieved, but NOx concentrations increase up to 78 ppm (Figure 8b). Moreover, with the increase of O_2 enrichment in air, NOx increases gradually, even an N_2 concentration in air decreases. This tendency was also determined with LCV gases containing higher LHV values, and the highest NOx concentrations were determined (up to 1100 ppm) burning the LCV gas with CH_4-25%. However, minor emissions of NOx (up to 20 and 40 ppm) were observed burning LCV gases with CH_4 content of 5 vol% and 10 vol%, respectively. Besides, the results show that during the combustion of waste biogas (up to 20 vol% CH_4) under oxygen-enriched conditions (25 and 30 vol% of O_2), the NOx and CO emissions do not exceed the established emission limits [30]. But the obtained results during the combustion of mixtures with a higher amount of CH_4 in CO_2 confirm that the NOx formation strongly belongs to the thermal NO formation route under oxygen-enriched conditions. As it mainly depends on the flame temperature, the O_2 enrichment level should be controlled based on the LHV value to prevent unwanted NOx formation.

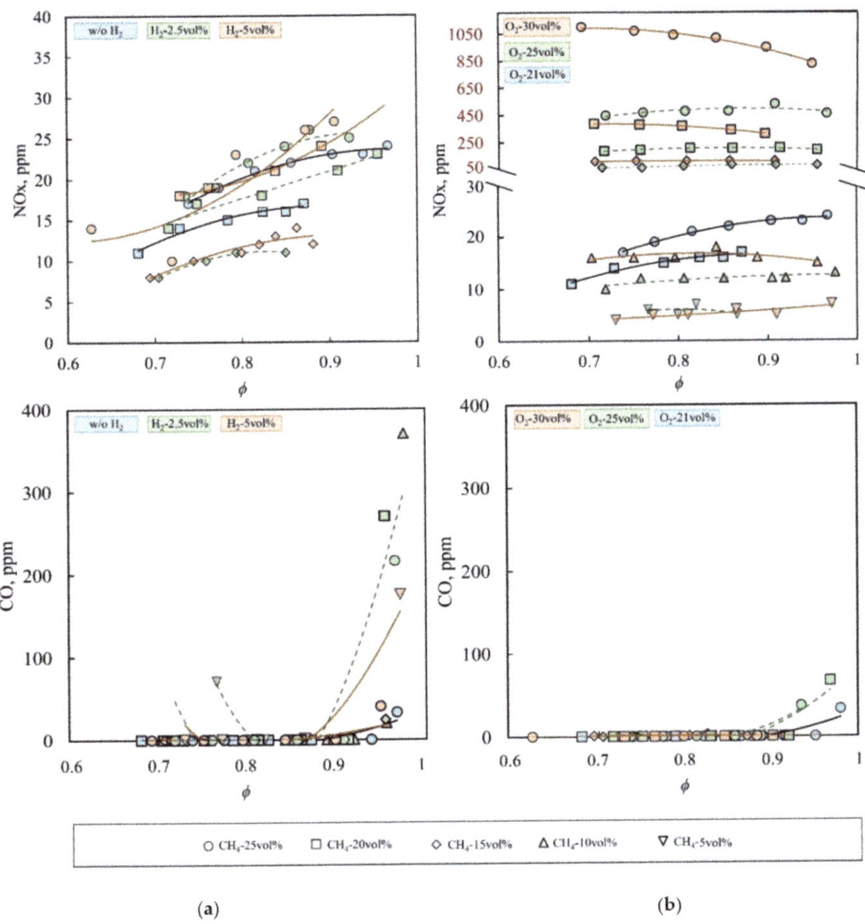

Figure 8. NOx and CO emissions of all tested waste biogas (LCV) mixture combustion under hydrogen-enrichment (**a**) and oxygen-enriched air (**b**) versus fuel-to-air ratio.

4. Conclusions

This study describes a flexible swirl burner (~100 kW) with an adiabatic chamber, which was designed and developed to burn unstable composition waste biogases (5–30 vol% of CH_4 in CO_2) and performed tests at real conditions. During the experimental tests, the flame stability and combustion limits were explored under air, oxygen-enriched atmosphere, or with hydrogen addition. The stable operation of non-premixed swirl combustion of waste biogas in the air was achieved for CH_4 content in CO_2 with not less than 20 vol%. Further decrease of CH_4 content in CO_2 requires H_2 or O_2 enrichment to extend flammability and improve flame stability. It was determined that by adding up to 5 vol% of H_2, the stable combustion of waste biogas with CH_4 content of 15 vol% is obtained. A greater improvement in flame stability of waste biogas was established using O_2-enrichment. The supply of O_2 by 25 and 30 vol% resulted in the stable combustion of waste biogases gases containing 10 vol% and

5 vol% of CH_4, respectively. Besides, the tests of the developed swirl burner at real conditions revealed that the flame blow-off is more likely at higher fuel to air ratios ($\phi > 0.95$), and this was established for all studied cases.

This work shows the different stabilization modes and the changes in intensities of OH*chemiluminescence spectra, which enables to determine the possible transition between regimes and when flammability limit might occur. The linear correlation between the spectral intensity ratio of OH* and CH* (I_{OH*}/I_{CH*}) and CH_4 content in LCV gas is determined, which clearly demonstrates the change in intensity for the tested modes and can be used for the prediction of blow-off limits burning different LCV gases under different H_2 or O_2 enrichments.

During all tests, the online flue gas analysis was performed to determine the behavior of NOx and CO emissions and explore its relation to flame stability. Minor emissions of NOx (up to 20 and 40 ppm) are observed only burning waste biogas (LCV) in air and with H_2 addition. During combustion under oxygen-enriched conditions, such concentrations can be obtained only by combusting waste biogases containing the CH_4 content of 5 vol% and 10 vol%, respectively. The CO emissions in most cases, were determined to be at zero levels and formed only by approaching the blow-off limit. In that case, the proposed combustion system meets the EU established emission standards for gas-burning systems.

However, if waste biogas calorific value increase, O_2 enrichment should be minimized to prevent unwanted high NOx formation. Besides, thermal damage to the combustion system and high NOx emissions are possible if the oxygen-enriched air is supplied with higher O_2 concentrations than 30 vol%.

Overall, the developed swirl burner with the adiabatic combustion chamber is a flexible combustion system that could be installed in the biogas plant to produce heat and electricity and also provides a solution to recover primary energy and reduce pollutant emissions from biomethane production obtained waste biogases.

Author Contributions: Conceptualization, N.S. and R.P.; methodology, N.S.; software, R.S.; validation, N.S. and R.S.; formal analysis, N.S. and R.P.; investigation, N.S., R.P., R.S. and A.L.; resources, N.S.; data curation, N.S. and R.P.; writing—original draft preparation, N.S. and R.P.; writing—review and editing, R.S. and A.L.; visualization, A.L.; supervision, N.S.; project administration, N.S.; funding acquisition, R.S. All authors have read and agreed to the published version of the manuscript.

Funding: This project has received funding from the Research Council of Lithuania (LMTLT), agreement No P-MIP-17-26.

Conflicts of Interest: The authors declare no conflict of interest.

References

1. Rekleitis, G.; Haralambous, K.-J.; Loizidou, M.; Aravossis, K. Utilization of Agricultural and Livestock Waste in Anaerobic Digestion (A.D): Applying the Biorefinery Concept in a Circular Economy. *Energies* **2020**, *13*, 4428. [CrossRef]
2. Stančin, H.; Mikulčić, H.; Wang, X.; Duić, N. A review on alternative fuels in future energy system. *Renew. Sustain. Energy Rev.* **2020**, *128*, 109927. [CrossRef]
3. Kumar, G.; Kim, S.-H.; Lay, C.-H.; Ponnusamy, V.K. Recent developments on alternative fuels, energy and environment for sustainability. *Bioresour. Technol.* **2020**, *317*, 124010. [CrossRef] [PubMed]
4. Tanaka, R.; Shinoda, M.; Arai, N. Combustion characteristics of a heat-recirculating ceramic burner using a low-calorific-fuel. *Energy Convers. Manag.* **2001**, *42*, 1897–1907. [CrossRef]
5. Nguyen, P.D.; Ghazal, G.; Piñera, V.C.; Battaglia, V.; Rensgard, A.; Ekman, T.; Gazdallah, M. Modelling of flameless oxy-fuel combustion with emphasis on radiative heat transfer for low calorific value blast furnace gas. *Energy Procedia* **2017**, *120*, 492–499. [CrossRef]
6. Hosseini, S.E.; Wahid, M.A. Biogas utilization: Experimental investigation on biogas flameless combustion in lab-scale furnace. *Energy Convers. Manag.* **2013**, *74*, 426–432. [CrossRef]
7. Wang, G.; Tang, P.; Li, Y.; Xu, J.; Durst, F. Flame front stability of low calorific fuel gas combustion with preheated air in a porous burner. *Energy* **2019**, *170*, 1279–1288. [CrossRef]

8. Kiedrzyńska, A.; Lewtak, R.; Świątkowski, B.; Jóźwiak, P.; Hercog, J.; Badyda, K. Numerical study of natural gas and low-calorific syngas Co-Firing in a pilot scale burner. *Energy* **2020**, 118552. [CrossRef]

9. Al-Halbouni, A.; Rahms, H.; Görner, K. An efficient combustion concept for low calorific gases. *Renew. Energy Power Qual. J.* **2007**, *1*, 45–50. [CrossRef]

10. Mörtberg, M.; Blasiak, W.; Gupta, A.K. Combustion of normal and low calorific fuels in high temperature and oxygen deficient environment. *Combust. Sci. Technol.* **2006**, *178*, 1345–1372. [CrossRef]

11. Colorado, A.; McDonell, V. Surface stabilized combustion technology: An experimental evaluation of the extent of its fuel-flexibility and pollutant emissions using low and high calorific value fuels. *Appl. Therm. Eng.* **2018**, *136*, 206–218. [CrossRef]

12. Al-Hamamre, Z.; Diezinger, S.; Talukdar, P.; Von Issendorff, F.; Trimis, D. Combustion of Low Calorific Gases from Landfills and Waste Pyrolysis Using Porous Medium Burner Technology. *Process. Saf. Environ. Prot.* **2006**, *84*, 297–308. [CrossRef]

13. Al-attab, K.A.; Ho, J.C.; Zainal, Z.A. Experimental investigation of submerged flame in packed bed porous media burner fueled by low heating value producer gas. *Exp. Therm. Fluid Sci.* **2015**, *62*, 1–8. [CrossRef]

14. Song, F.; Wen, Z.; Dong, Z.; Wang, E.; Liu, X. Ultra-Low calorific gas combustion in a gradually-varied porous burner with annular heat recirculation. *Energy* **2017**, *119*, 497–503. [CrossRef]

15. Wood, S.; Harris, A.T. Porous burners for lean-burn applications. *Prog. Energy Combust. Sci.* **2008**, *34*, 667–684. [CrossRef]

16. Wei, Z.L.; Leung, C.W.; Cheung, C.S.; Huang, Z.H. Effects of equivalence ratio, H_2 and CO_2 addition on the heat release characteristics of premixed laminar biogas-hydrogen flame. *Int. J. Hydrog. Energy* **2016**, *41*, 6567–6580. [CrossRef]

17. Leung, T.; Wierzba, I. The effect of hydrogen addition on biogas non-premixed jet flame stability in a co-flowing air stream. *Int. J. Hydrog. Energy* **2008**, *33*, 3856–3862. [CrossRef]

18. García-Armingol, T.; Ballester, J. Flame chemiluminescence in premixed combustion of hydrogen-enriched fuels. *Int. J. Hydrog. Energy* **2014**, *39*, 11299–11307. [CrossRef]

19. García-Armingol, T.; Ballester, J. Operational issues in premixed combustion of hydrogen-enriched and syngas fuels. *Int. J. Hydrog. Energy* **2015**, *40*, 1229–1243. [CrossRef]

20. Bâ, A.; Cessou, A.; Marcano, N.; Panier, F.; Tsiava, R.; Cassarino, G.; Ferrand, L.; Honoré, D. Oxyfuel combustion and reactants preheating to enhance turbulent flame stabilization of low calorific blast furnace gas. *Fuel* **2019**, *242*, 211–221. [CrossRef]

21. Khalil, A.E.; Gupta, A.K. Flame fluctuations in Oxy-CO_2-methane mixtures in swirl assisted distributed combustion. *Appl. Energy* **2017**, *204*, 303–317. [CrossRef]

22. Francisco, R.W.; Rua, F.; Costa, M.; Catapan, R.C.; Oliveira, A.A.M. On the Combustion of Hydrogen-Rich Gaseous Fuels with Low Calorific Value in a Porous Burner. *Energy Fuels* **2010**, *24*, 880–887. [CrossRef]

23. Chiu, C.-P.; Yeh, S.-I.; Tsai, Y.-C.; Yang, J.-T. An Investigation of Fuel Mixing and Reaction in a CH_4/Syngas/Air Premixed Impinging Flame with Varied H_2/CO Proportion. *Energies* **2017**, *10*, 900. [CrossRef]

24. Karyeyen, S.; Ilbas, M. Experimental and numerical analysis of turbulent premixed combustion of low calorific value coal gases in a generated premixed burner. *Fuel* **2018**, *220*, 586–598. [CrossRef]

25. Ilbas, M.; Bektaş, A.; Karyeyen, S. A new burner for oxy-fuel combustion of hydrogen containing low-calorific value syngases: An experimental and numerical study. *Fuel* **2019**, *256*, 115990. [CrossRef]

26. Yilmaz, İ.; Alabaş, B.; Taştan, M.; Tunç, G. Effect of oxygen enrichment on the flame stability and emissions during biogas combustion: An experimental study. *Fuel* **2020**, *280*, 118703. [CrossRef]

27. Striūgas, N.; Zakarauskas, K.; Paulauskas, R.; Skvorčinskienė, R. Chemiluminescence-Based characterization of tail biogas combustion stability under syngas and oxygen-enriched conditions. *Exp. Therm. Fluid Sci.* **2020**, *116*, 110133. [CrossRef]

28. Nemitallah, M.; AlKhaldi, S.; Abdelhafez, A.; Habib, M. Effect analysis on the macrostructure and static stability limits of oxy-methane flames in a premixed swirl combustor. *Energy* **2018**, *159*, 86–96. [CrossRef]

29. Striūgas, N.; Tamošiūnas, A.; Marcinauskas, L.; Paulauskas, R.; Zakarauskas, K.; Skvorčinskienė, R. A sustainable approach for plasma reforming of tail biogas for onsite syngas production during lean combustion operation. *Energy Convers. Manag.* **2020**, *209*, 112617. [CrossRef]

30. European Parliament, Council of the European Union. Directive (EU) 2015/2193 of the European Parliament and of the Council on the limitation of emissions of certain pollutants into the air from medium combustion plants. *Off. J. Eur. Union EN* **2014**, *451*, 123–134.

31. Ballester, J.; García-Armingol, T. Diagnostic techniques for the monitoring and control of practical flames. *Prog. Energy Combust. Sci.* **2010**, *36*, 375–411. [CrossRef]

32. Devriendt, K.; Van Look, H.; Ceursters, B.; Peeters, J. Kinetics of formation of chemiluminescent CH(A2Δ) by the elementary reactions of $C_2H(X2\Sigma+)$ with O(3P) and $O_2(X3\Sigma g-)$: A pulse laser photolysis study. *Chem. Phys. Lett.* **1996**, *261*, 450–456. [CrossRef]

33. Kojima, J.; Ikeda, Y.; Nakajima, T. Basic aspects of OH(A), CH(A), and C2(d) chemiluminescence in the reaction zone of laminar methane–air premixed flames. *Combust. Flame* **2005**, *140*, 34–45. [CrossRef]

34. Quintino, F.M.; Trindade, T.P.; Fernandes, E.C. Biogas combustion: Chemiluminescence fingerprint. *Fuel* **2018**, *231*, 328–340. [CrossRef]

35. Zhen, H.S.; Leung, C.W.; Cheung, C.S. Effects of hydrogen addition on the characteristics of a biogas diffusion flame. *Int. J. Hydrog. Energy* **2013**, *38*, 6874–6881. [CrossRef]

36. Zhen, H.S.; Leung, C.W.; Cheung, C.S.; Huang, Z.H. Characterization of biogas-hydrogen premixed flames using Bunsen burner. *Int. J. Hydrog. Energy* **2014**, *39*, 13292–13299. [CrossRef]

37. Song, Y.; Zou, C.; He, Y.; Zheng, C. The chemical mechanism of the effect of CO_2 on the temperature in methane oxy-fuel combustion. *Int. J. Heat Mass Transf.* **2015**, *86*, 622–628. [CrossRef]

38. Guiberti, T.F.; Durox, D.; Schuller, T. Flame chemiluminescence from CO_2- and N2-diluted laminar CH_4/air premixed flames. *Combust. Flame* **2017**, *181*, 110–122. [CrossRef]

39. Paulauskas, R.; Martuzevičius, D.; Patel, R.B.; Pelders, J.E.H.; Nijdam, S.; Dam, N.J.; Tichonovas, M.; Striūgas, N.; Zakarauskas, K. Biogas combustion with various oxidizers in a nanosecond DBD microplasma burner. *Exp. Therm. Fluid Sci.* **2020**, *118*, 110166. [CrossRef]

40. Oh, J. Spectral characteristics of a premixed oxy-methane flame in atmospheric conditions. *Energy* **2016**, *116*, 986–997. [CrossRef]

41. Yang, J.; Gong, Y.; Guo, Q.; Zhu, H.; Wang, F.; Yu, G. Experimental studies of the effects of global equivalence ratio and CO_2 dilution level on the OH* and CH* chemiluminescence in CH_4/O_2 diffusion flames. *Fuel* **2020**, *278*, 118307. [CrossRef]

Article

Efficiency Analysis and Integrated Design of Rocket-Augmented Turbine-Based Combined Cycle Engines with Trajectory Optimization

Feng Guo [1], Wenguo Luo [1], Feng Gui [2], Jianfeng Zhu [1,*], Yancheng You [1] and Fei Xing [1]

[1] School of Aerospace Engineering, Xiamen University, Xiamen 361005, China;
34720170154965@stu.xmu.edu.cn (F.G.); 34720181152195@stu.xmu.edu.cn (W.L.);
yancheng.you@xmu.edu.cn (Y.Y.); f.xing@xmu.edu.cn (F.X.)

[2] AECC Sichuan Gas Turbine Establishment, Chengdu 610500, China; gflysky@126.com

* Correspondence: zhjf@xmu.edu.cn

Received: 9 May 2020; Accepted: 2 June 2020; Published: 5 June 2020

Abstract: An integrated analysis method for a rocket-augmented turbine-based combined cycle (TBCC) engine is proposed based on the trajectory optimization method of the Gauss pseudospectral. The efficiency and energy of the vehicles with and without the rocket are analyzed. Introducing an appropriate rocket to assist the TBCC-powered vehicle will reduce the total energy consumption of drag, and increase the vehicle efficiency in the transonic and the mode transition. It results in an increase in the total efficiency despite a reduction in engine efficiency. Therefore, introducing a rocket as the auxiliary power is not only a practical solution to enable flight over a wide-speed range when the TBCC is incapable but also probably an economical scheme when the the TBCC meets the requirements of thrust. When the vehicle drag is low, the rocket works for a short time and its optimal relative thrust is small. Thus, the TBCC combined with a booster rocket will be a more simple and suitable scheme. When the vehicle drag is high, the operating time of the rocket is long and the optimal relative thrust is large. The specific impulse has a significant impact on the flight time and the total fuel consumption. Accordingly, the combination form for the rocket-based combined cycle (RBCC) engines and the turbine will be more appropriate to obtain higher economic performance.

Keywords: turbine-based combined cycle engine; TBCC; rocket-augmented; trajectory optimization; Gauss pseudospectral method; efficiency analysis; combined design; integrated design

1. Introduction

Turbine-based combined cycle (TBCC) engines are believed to be a promising means of power for wide-speed range hypersonic vehicles [1]. Due to combining turbine engines with ramjet engines or dual-mode scramjet engines, typical TBCC engines have the advantages of low fuel consumption, the high reliability of turbine engines and the high-speed cruise ability of ramjet/dual mode ramjet (DMRJ) engines [2]. However, several technical issues with TBCC engines still exist, such as "Thrust Pinch", so the engines are incapable of providing sufficient thrust for acceleration during transonic and mode transition [3]. To resolve this problem, the "rocket-augmented TBCC" concept has been proposed as a near-term solution, where the rockets are used as auxiliary power for thrust augmentation when extra thrust is needed.

In terms of combination form, "rocket-augmented TBCC" engines are classified into two broad categories: "T+RBCC" and "R+TBCC". For "T+RBCC" engines, the rockets are typically mounted within the high-speed ducts and integrated as ejector-ramjets (ERJs). To date, some schemes have been published, such as TriJet [3,4], TRRE [5,6], and XTER [7]. All these engines utilize the thrust and specific impulse potential of ERJs [8–10], but the rockets should be synergistically integrated into the thermodynamic cycle at the cost of complexity of the rocket-matching design. By comparison, the rockets in "R+TBCC" engines are installed as additional boosters, which are generally isolated from the typical TBCC engines. Many researchers have discussed these kinds of engine and vehicle as well, for example, the advanced Rocket/Dual-mode ramjet propulsion of Lapcat II [11] and the high-speed near space flight vehicle proposed by the Beijing Institute of Mechanical and Electrical Engineering [12]. Although the integration design method for the "R+TBCC" engines is relatively simple, this combination may increase extra vehicle drag and inert weight, which may raise the fuel consumption and reduce the payload weight, respectively. From the above, in order to achieve synergistic benefits from the collaboration of TBCCs and rockets excellently, the performance of the vehicle, TBCC, and rockets should be taken into account when designing the combination form of the rocket-augmented TBCC.

Generally, the rockets are introduced to meet the vehicle thrust requirement when the TBCCs are incapable of providing sufficient thrust. Naturally, due to the low specific impulse (Isp) of rockets, the economic performance of the vehicle seems to be reduced. However, in flight, the acceleration increase by the rockets reduces the flight time and results in fuel saving for the TBCC engine. Therefore, the rockets used for the thrust augmentation have a mixed effect on the total economic performance. In addition, the acceleration characteristics are related to trajectory features; for example, the trajectory of the "climb-dive" (or gravity-assist) is used to overcome the transonic thrust pinch in the acceleration phase. Thus, the tradeoff should take into consideration the trajectory features in the thrust pinch regions. The rocket-matching analysis based on trajectory optimization is necessary for the integrated design of the rocket-augmented TBCC. Trajectory optimization methods have been studied for decades [13–15]. The Gauss pseudospectral method as a direct method has the ability to obtain accurate estimates of the state, costate, and control for continuous time optimal control problems [16] and has been successfully implemented in trajectory optimization problems of supersonic or hypersonic vehicles [17,18]. Thus, the Gauss pseudospectral method could be directly used to evaluate the integrated design of a rocket-augmented TBCC under different rocket schemes, and the results from trajectory optimization could be used for choosing the appropriate combination form for rockets and TBCCs for vehicle systems.

The paper is organized as follows: Section 2 will give a problem statement of the rocket-augmented strategy from the aspect of efficiency. In Section 3, an integrated analysis model for a rocket-augmented TBCC-powered vehicle with the trajectory optimization method of the Gaussian pseudospectral is established. Section 4 analyzes the performance of the vehicle from the aspect of efficiency and energy, and the rocket thrust is optimized in a specified vehicle case. Based on the above understanding, the parameter study of different vehicles and rockets is discussed in Section 5, and the combination form for rockets and TBCCs is given. Finally, Section 6 summarizes this paper.

2. Problem Statement

TBCC vehicles suffer from two main thrust pinches [1], as shown in Figure 1a. One is the transonic thrust pinch, and the other is the transition thrust pinch. The thrust pinches are limited by the current performance level of vehicle and TBCC engines. The thrust pinch issues can be described as flight with small acceleration and long acceleration times, which can be measured by vehicle efficiency, η_{veh} [18]:

$$\eta_{veh} = E_{veh}/(TV cos\alpha) \tag{1}$$

E_{veh} represents the vehicle work:

$$E_{veh} = (Tcos\alpha - D - mgsinr)V \tag{2}$$

where T and D denote the thrust and drag, respectively. α and r denote angle of attack and the flight path angle, respectively. m denotes the mass of the vehicle, and V denotes the flight velocity.

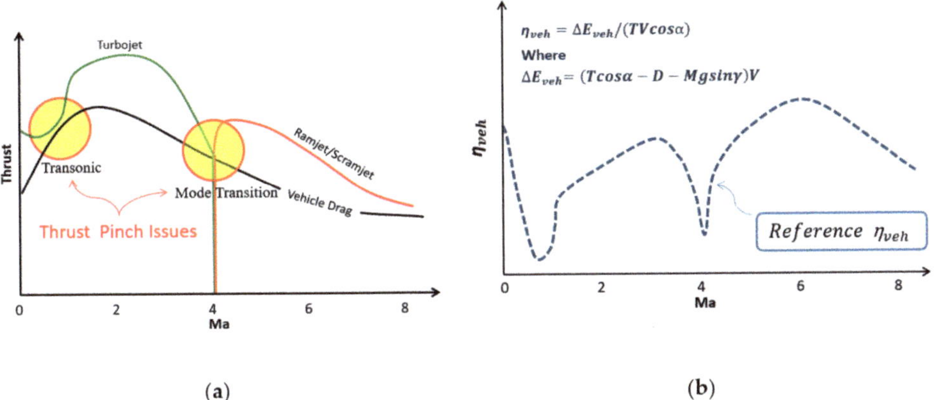

(**a**)   (**b**)

Figure 1. (**a**) Thrust pinch in transonic and mode transition; (**b**) Vehicle efficiency.

As shown in Figure 1b, the reference η_{veh} curve could effectively reflect the acceleration efficiency characteristics corresponding to Figure 1a. It can be seen that the vehicle efficiency in the thrust pinch region is very low.

For wide-speed flight, the TBCC has the distinct advantage of utilizing oxygen in the air instead of carrying it onboard, like a rocket [19]. This results in the higher Isp of TBCC engines compared to rockets, as shown in Figure 2a [20]. Similarly, the engine efficiency η_{enigne} can be formulated with the Isp [21]:

$$\eta_{engine} = \frac{gV}{h_{PR}} \cdot Isp \tag{3}$$

where h_{PR} denotes the heating value of the fuel.

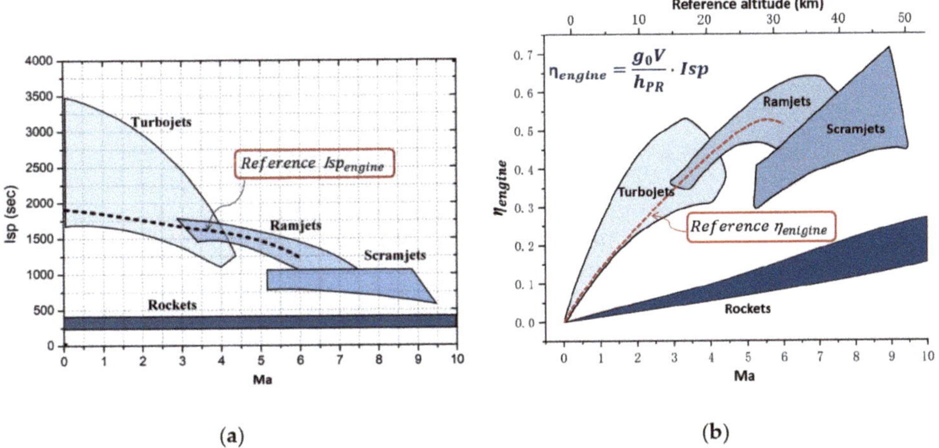

(**a**)   (**b**)

Figure 2. (**a**) Reference Isp of engines; (**b**) Reference efficiency of engines.

As shown in Figure 2b, the rockets' engine efficiency is obviously lower than that of the air-breathing engines as well, such as turbojets, ramjets, and scramjets. However, rockets have significantly higher unit frontal area thrust (or unit weight thrust). Besides, the thrust of rockets increases gradually as the flight altitude increases, while the air-breathing engines have the opposite trend due to the reduction in air density. Therefore, a smaller rocket may generate enough thrust to accelerate the vehicle when going through the thrust pinch region, especially at high altitudes.

According to the description of Figures 1 and 2, introducing a small rocket to assist the TBCC could significantly increase the vehicle efficiency, η_{veh}, in the transonic and mode transition. Although the engine efficiency, η_{enigne}, may be reduced when the rocket works, the rocket-augmented strategy is still possible to improve the local total efficiency η_{tot}, as shown in Figure 3. It is to be noted that η_{tot} is defined as follows:

$$\eta_{tot} = \eta_{veh} \cdot \eta_{engine} \tag{4}$$

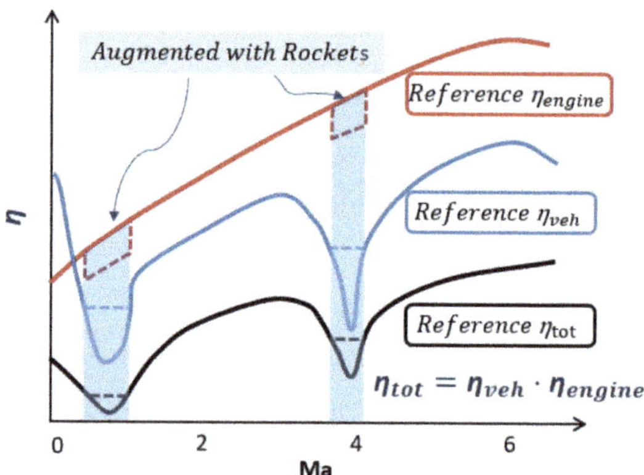

Figure 3. Total efficiency might be improved by rockets.

This analysis shows that introducing a rocket as the auxiliary power is not only a practical solution to enable flight over a wide-speed range when the TBCC is incapable but also probably an economical scheme when the TBCC meets the requirements of thrust.

3. Integrated Analysis Method

3.1. Description of the Vehicle Model

The baseline vehicle is assumed to have a take-off weight of 19,050 kg and a wing loading of 460 kg/m². As a result of a lack of performance data for currently available TBCC-powered vehicles, the aerodynamic coefficients in high Mach (>1.5) refer to X − 43 [22], and the data for low Mach (≤1.5) are obtained from the reference [23]. It is noted that the low-Mach data in the case of X − 43 are not suitable for showing the transonic thrust pinch.

For the trajectory optimization in the conceptual design, a point mass model for motion in a vertical plane is usually quite adequate [23].

The equations of motion for this model [24] are given by:

$$\dot{h} = V sin\gamma \tag{5}$$

$$\dot{V} = \frac{Tcos\alpha \; - \; D}{m} - gsin\gamma \tag{6}$$

$$\dot{\gamma} = \frac{Tsin\alpha + L}{mV} + \left(\frac{V}{r} - \frac{g}{V}\right)cos\gamma \tag{7}$$

$$\dot{m} = -\frac{T}{g \cdot Isp} \tag{8}$$

The lift L and drag D are defined as:

$$L = qSC_L(\alpha, Ma) \tag{9}$$

$$D = qSC_D(\alpha, Ma) \tag{10}$$

where q is the flight dynamic pressure and S is the reference area of the vehicle. The lift coefficient C_L and the drag coefficient C_D are the interpolation functions of the attack angle α and the Mach number.

3.2. Description of the Engines Model

The TBCC engine model adopts the lapping combination of the Ma 4.4 turbine and the HRE scramjet from the Air Force Research Laboratory (AFRL) [22]. The ratio of the number of turbine engines to ramjet engines is 20:1, and the takeoff thrust/weight ratio is 1.0. The thrust and the Isp are obtained by interpolating the altitude and the velocity data table. To highlight the "thrust trap" feature of the mode transition, the thrust and the Isp are calculated by Equations (11) and (12) [25], respectively:

$$T(\lambda, M) = T_{turb}\left(\frac{M_2 - M}{M_2 - M_1}\right)^{\lambda} + T_{ram}\left(\frac{M - M_1}{M_2 - M_1}\right)^{\lambda}, M \in [M_1, M_2] \tag{11}$$

$$Isp(\lambda, M) = \frac{T(\lambda, M)}{\left(\frac{T_{turb}}{Isp_{turb}}\left(\frac{M_2 - M}{M_2 - M_1}\right)^{\lambda} + \frac{T_{ram}}{Isp_{ram}}\left(\frac{M - M_1}{M_2 - M_1}\right)^{\lambda}\right)}, M \in [M_1, M_2] \tag{12}$$

where M_1 (4.0) and M_2 (4.4) are the start and the end Mach numbers of the mode transition, and the parameter λ is the pinch coefficient and can measure the minimum thrust in mode transition. In this paper, the $T\left(\lambda, \frac{M_1 + M_2}{2}\right)$ is equal to the 2/3 of $T\left(1, \frac{M_1 + M_2}{2}\right)$, in which the λ is correspondingly given as 1.585.

The Isp of the baseline rocket is 300 s. The thrust of the baseline rocket is adjustable, and the adjustable thrust is the product of the rocket throttle, Thr, and the maximum of thrust, T_{Rocket_max}:

$$T_{Rocket} = T_{Rocket\_max} \cdot Thr \tag{13}$$

where T_{Rocket_max} is defined as the product of the vehicle takeoff weight, m_{to}, and the maximum relative thrust of the rocket, RT_{max}:

$$T_{Rocket\_max} = m_{to} \cdot g \cdot RT_{max} \tag{14}$$

In addition, introducing the rocket increases the frontal area of the vehicle and the weight of the propulsion system, which influences the performance of the vehicle. In this study, the rocket weight is estimated at 1/50 of the maximum thrust. The drag coefficient is used to estimate the performance affected by the frontal area [12].

3.3. Integrated Trajectory Optimization Method

3.3.1. Optimization Problem

According to both the vehicle model and the engine model, the flight altitude h, speed V, flight path angle γ, vehicle mass m, and attack angle α are set as state parameters. Meanwhile, the gradient of the attack angle $\dot{\alpha}$ and rocket throttle ratio thr are set as control variables. It is noted that the attack angle $\dot{\alpha}$ is aimed at smoothing flight control. Based on the vehicle control equations, the optimization problem of the ascent trajectory is formulated with the corresponding cost function and parameter constraints.

The objective is to determine the minimum-fuel (minF) trajectory and control from takeoff to a specified speed and altitude. Besides, the minimum-time (minT) trajectory optimization problem is analyzed as a comparative case. The cost functions are formulated as

$$J_{min,fuel} = -m(t_f)$$ (15)

$$J_{min,time} = t_f$$ (16)

The boundary conditions and constraints are shown in Table 1.

Table 1. Boundary conditions and constraints.

Value Type	Time	Control Parameter		State Parameter				
	t/s	$\dot{\alpha}$(deg/s)	thr	h/km	V/(m/s)	γ/deg	m/kg	α/deg
Initial	0	Free	Free	0	129.314	0	19,050.8	6
Terminal	Free	Free	Free	Free	1495.28	0	Free	Free
Min	0	−0.5	0	0	5	−40	22	−5
Max	1500	0.5	1	30	1794	40	19,050.8	20

Additionally, the operating range of dynamic pressures, q, is also restrained in different flight phases.

$$\begin{cases} 10kPa \leq q \leq 75kPa, & Ma \leq M_1, Ma > M_2 \\ 40kPa \leq q \leq 75kPa, & M_1 < Ma \leq M_2 \end{cases}$$ (17)

3.3.2. Gauss Pseudospectral Method

The optimization problem above can be converted to a continuous Bolza problem by using the Gauss pseudospectral method (GPM). The GPM constructs the Lagrange interpolation polynomials on a set of Legendre points to approximate the state variables and control variables of the system, so the continuous-optimal control problem can be transcribed into a nonlinear planning problem (NLP). Studies have shown that Karush–Kuhn–Tucker (KKT) multipliers of the NLP can be utilized to estimate the costate at both the Legendre-Gauss points and the boundary points accurately, which is due to the equivalence between the KKT conditions and the discretized first-order necessary conditions [12]. In this paper, the trajectory optimization problem is solved by the general software package GPOPS; more details of this algorithm are discussed in [26,27].

4. Efficiency Analysis and Rocket Optimization

According to the predictions in the problem statement, the engine efficiency is depressed when the rocket is introduced, yet the vehicle efficiency in thrust pinches may be improved significantly. With overall consideration, introducing a rocket is still possible to increase the total efficiency. In this section, the efficiency analysis of the trajectory with the rocket is performed on the baseline TBCC-powered vehicle by using the integrated analysis method. However, introducing a rocket increases the frontal area of the vehicle and the weight of the propulsion system. Taking the above factor into account, the rocket with an appropriate thrust will yield the optimal performance. Therefore, the rocket optimization is also studied in this section.

Figure 4 shows the trajectories of the baseline vehicle with minF and minT as the optimization objectives, and of the rocket-augmented TBCC-powered vehicle with minF as the optimization objective. Simulations are performed from Mach 0.38 to 5 with the constraints given in Table 1. For the baseline vehicle, the minF path shows that the vehicle goes through transonic and mode transition in the way of rapid climb and dive, and flies at the constant dynamic pressure of 75 kPa in most of the remaining voyage. By contrast, the minT path shows a relatively lower climb altitude during the transonic and without "climb-dive" during mode transition. It is indicated that the "climb-dive" trajectory strategy is an efficient means of fuel saving in the thrust pinch regions. It is also necessary when going through transonic in the path of minT, which has been discussed in J. Zheng et al.'s work [25].

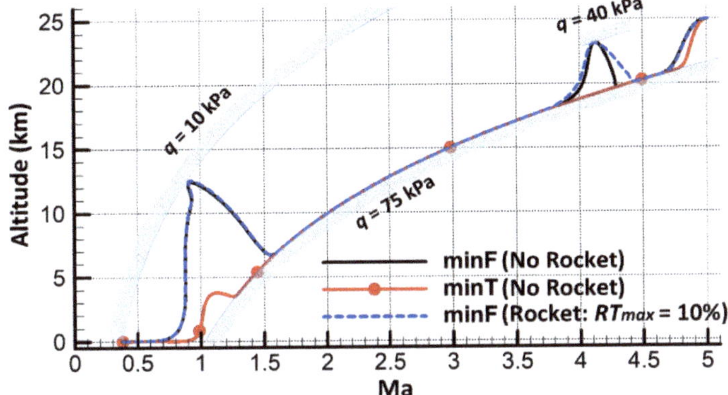

Figure 4. Flight envelopes.

Based on the TBCC model, the rocket with an RT_{max} of 10% is used to optimize the trajectory of minF, shown as the dashed line in Figure 4. Interestingly, the rocket does not work in the transonic region, but only turns on during the mode transition. Moreover, the vehicle firstly climbs to the limit boundary of the minimum dynamic pressure of 40 kPa and then takes advantage of the gravity-assist to accelerate during the mode transition. Due to the change in trajectory when the rocket works, the local performance of vehicle might be changed.

Figure 5 shows a detailed comparison of the flight time of the three trajectories above. For the TBCC-powered vehicle, the times taken to reach Mach 5 of minF and minT are 719.2 s and 614.5 s, respectively. The time difference between the minF and minT closely depends on the different trajectory strategy in the transonic. The "climb-dive" trajectory strategy for fuel saving takes a great amount of time. The total time of the rocket-augmented minF is 648.4 s, which is 9.8% less than that of the baseline minF. The time difference mainly depends on the different accelerations in the mode transition. In addition, as a result of the flight drag caused by the rocket, the trajectory of the minF with rocket takes slightly more flight time than the baseline minF to accelerate to the same Mach number.

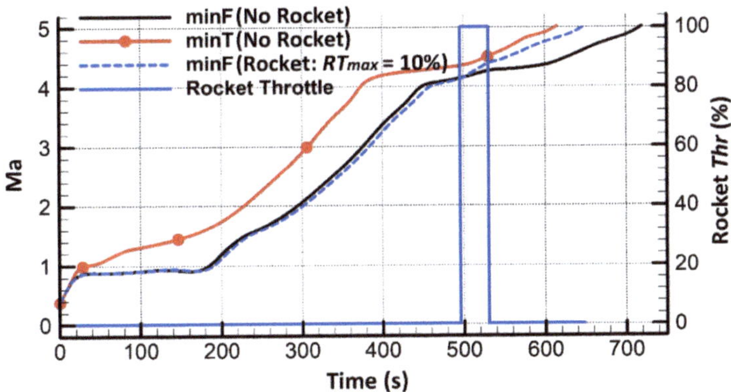

Figure 5. Flight time profiles.

The throttle control of the rocket is also shown in Figure 5. The rocket works at 495.8 s and lasts for 34.3 s. The corresponding Mach number ranges from 4.12 to 4.39. By contrast, the TBCC-powered vehicle without a rocket takes 119.5 s to flight over the same Mach range. In addition, the rocket throttle is allowed to vary freely between 0% and 100% in the trajectory optimization; however, the results show that the throttle mostly remains in the two states of 0% and 100%. This indicates that the adjustment of thrust is not necessary. Therefore, the fixed-thrust design could satisfy the thrust requirement and simplify the structure of the rocket.

Figure 6 shows the vehicle mass variations with the flight Mach. The total fuel consumption of the minT, the minF, and the rocket-augmented minF to reach Mach 5 are 4992.8 kg, 4568.4 kg, and 4444.2 kg, respectively. The fuel saving is 424.4 kg (8.5%) from the minT to the minF, and a further 124.2 kg (2.7%) from the minF to the rocket-augmented minF. Due to the gravity-assist, compared with the minT, the minF saves 415.3 kg of fuel in the process of transonic, contributing 97.9% of the total fuel saving. With the assistance of the rocket, an additional 201.6 kg of fuel is saved during the mode transition, which improves the economic performance of the TBCC vehicle even with the additional fuel consumption resulting from the drag of the rocket. If the weight of the rocket is considered as a part of the fuel consumption, the fuel saving is 86.1 kg (1.9%). Based on these results, it appears that the rocket-augmented TBCC is a better option to promote economic performance than the TBCC without a rocket for wide-speed range vehicles.

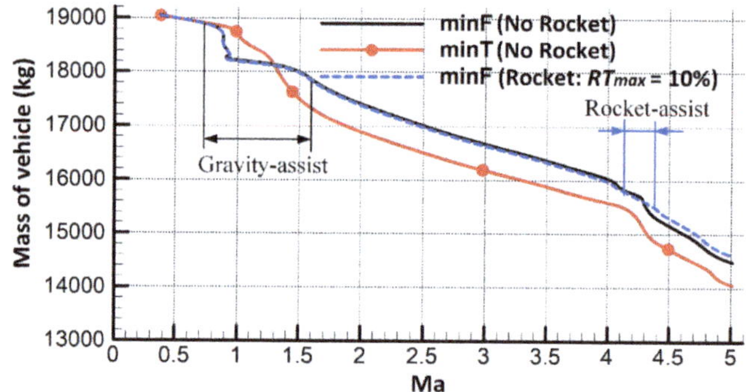

Figure 6. Vehicle mass profiles.

The TBCC-powered vehicle assisted by a rocket could improve both the flight timeliness and the economic performance, which could be explained in terms of the energy and efficiency.

From the perspective of energy, the vehicle climbs from a low-potential and low-kinetic state to a high-potential and high-kinetic state. In the process of ascending, besides increasing kinetic energy and potential energy, the thrust provides the most of the energy to overcome the drag. For instance, the drag dissipates 35.45 GJ of energy in the trajectory of the minT, while the increases in the kinetic energy and the potential energy are 15.56 GJ and 3.42 GJ, respectively. By contrast, the drag dissipates 31.31 GJ of energy in the trajectory of the minF, which is 11.7% less than that of the minT. Under the assistance of the rocket in the minF, the energy consumption of drag is reduced from 31.31 GJ to 26.58 GJ. The difference in the drag dissipation between the minF and the minT mainly lies in the transonic region, as shown in Figure 7a. The energy of the drag has a significant reduction from the minT to the minF, which results from the gravity-assist strategy. After climbing to a higher altitude, the vehicle flies with a smaller drag and lower energy consumption. Meanwhile, the thrust is reduced and the Isp of the engine remains almost constant in the process of ascending, which is one of the reasons for the fuel saving. Since the gravitational potential energy is comparable to the kinetic energy in the transonic region, the gravity-assist strategy could achieve an efficient increase in the kinetic energy from the conversion of the potential energy. However, in the mode transition, the potential energy is much less than the kinetic energy and the gravity-assist strategy is inefficient, as shown in Figure 7b. With the assistance of the rocket, the time spent in the mode transition is reduced from 164.5 s to 73.2 s, and the energy consumption of drag is reduced from 9.73 GJ to 3.47 GJ. The results indicate that the rocket-augmented scheme seems to be a better choice for the reduction of drag dissipation in mode transition.

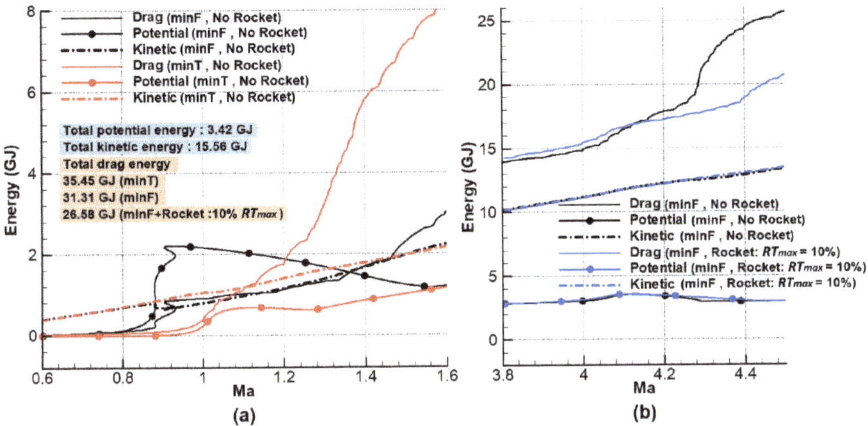

Figure 7. (**a**) Energy profiles in low Mach; (**b**) Energy profiles in high Mach.

Figure 8 shows the efficiency profiles of the two trajectories of the minF and the rocket-augmented minF. As shown in Figure 8a, the vehicle efficiency, η_{veh}, has a sharp decrease before Mach 0.9 due to the rapid dropping of acceleration in the phase of the climb. In the final stage of climb, the thrust is not able to accelerate the vehicle, so the value of η_{veh} is negative. With the rapid dive in transonic, the η_{veh} increases first and then decreases, under the combined effect of drag, thrust, and gravity. By contrast, the variation of the η_{enigne} is slight, because the Isp is insensitive to altitude.

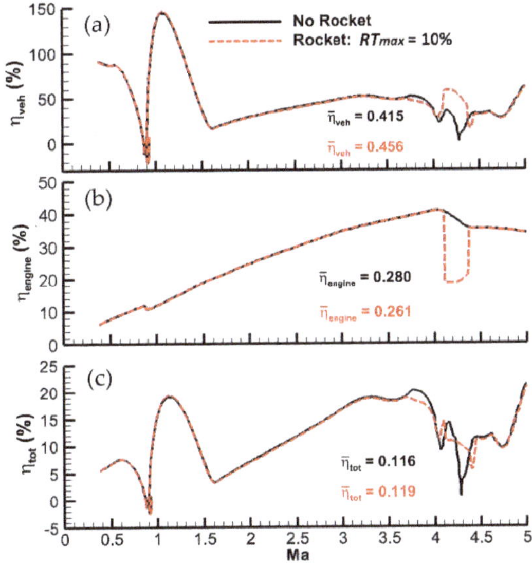

Figure 8. (**a**) Vehicle efficiency profiles; (**b**) Engine efficiency profiles; (**c**) Total efficiency profiles.

However, when the rocket works, the η_{veh} is improved and the η_{enigne} is reduced during the mode transition. Additionally, the η_{tot} is optimized, especially in the excessively inefficient region around Ma 4.3. In order to evaluate the effectiveness of the performance optimization, the average vehicle efficiency, engine efficiency, and total efficiency are respectively defined as:

$$\overline{\eta}_{veh} = \frac{\int \Delta E_{veh} dt}{\int TV\cos\alpha\, dt} \tag{18}$$

$$\overline{\eta}_{engine} = \frac{\int TV\cos\alpha\, dt}{\int h_{PR}\dot{m}_f dt} \tag{19}$$

$$\overline{\eta}_{tot} = \frac{\int \Delta E_{veh} dt}{\int h_{PR}\dot{m}_f dt} \tag{20}$$

where $\dot{m}_f$ denotes the fuel mass flow rate.

These variables in Equations (18)–(20) are defined in Section 2. The $\overline{\eta}_{veh}$ increases by 9.9% (from 0.415 to 0.456) after the introduction of the rocket, and the $\overline{\eta}_{engine}$ is reduced by 6.8%, resulting in an increase of 2.6% in the $\overline{\eta}_{tot}$. Therefore, introducing a suitable rocket to assist the TBCC increases the vehicle efficiency in the mode transition, which results in an increase in the total efficiency despite the reduction in engine efficiency.

It seems that the greater the thrust of the rocket, the better the vehicle efficiency. However, the increase in rocket thrust will inevitably bring increased drag and additional weight. The two factors should be considered in the thrust design for the rocket for optimal performance.

In Figure 9, when the rocket thrust is relatively small ($RT_{max} < 6\%$), as the thrust of the rocket increases, the fuel consumption of the TBCC reduces rapidly. Even if the fuel consumption and the weight of the rocket are increased, the total weight of consumption is still reduced. It is noted that the total weight in Figure 9 is the sum of the fuel consumption of the TBCC, the fuel consumption of the rocket, and the weight of the rocket. As the thrust increases ($RT_{max} > 6\%$), the fuel consumption of the TBCC increases slowly, which is due to the increase in the drag. In the meantime, the weight and the fuel consumption of the rocket increase almost linearly. Consequently, the total weight of consumption is at a minimum value when the relative thrust of the rocket is about 6%, which is 2.2% less than the total weight consumed by the vehicle without the rocket.

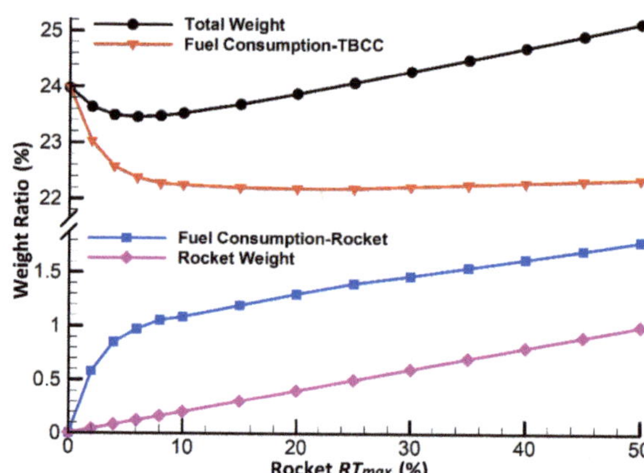

Figure 9. Rocket RT_{max} vs. weight consumption.

According to Equations (18) and (19), Figure 10 shows the variations in the average efficiencies with the rocket thrust. With the rocket RT_{max} increasing from 0% to 8%, the $\overline{\eta}_{veh}$ has a sharp increase (9.6%) while the $\overline{\eta}_{engine}$ drops by about 6%, which results in a 3% increase in the $\overline{\eta}_{tot}$. With the rocket RT_{max} further increasing ($RT_{max} \geq 8\%$), the $\overline{\eta}_{veh}$ shows a slightly decreasing trend and the $\overline{\eta}_{engine}$ keeps decreasing, so the $\overline{\eta}_{tot}$ starts to decrease.

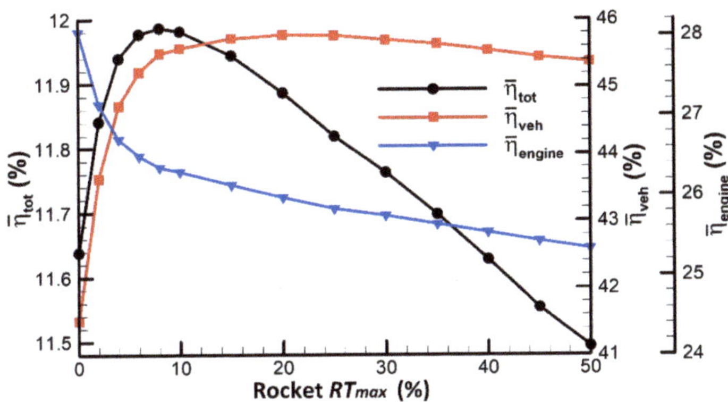

Figure 10. Rocket RT_{max} vs. average efficiencies.

With the assistance of the rocket, the time spent and the energy consumption of drag in the mode transition could be reduced significantly. Therefore, introducing an appropriate rocket to assist the TBCC-powered vehicle could increase vehicle efficiency in the mode transition, which results in an increase in total efficiency despite the reduction of engine efficiency. For a specified TBCC-powered vehicle, the thrust of the rocket has an optimal value to maximize the average total efficiency. The rocket with an overly small thrust could not improve the average vehicle efficiency sufficiently, but an overly large thrust might lead to excessive drag.

5. Parameter Study and Integrated Design

As mentioned previously, the performance of a specified TBCC-powered vehicle could be optimized by introducing the rocket. The essence of the optimization is the tradeoff between the efficiency and engine efficiency. Considering that vehicle drag and rocket Isp are respectively related to vehicle efficiency and engine efficiency, the optimization of the thrust and operating time of the rocket should involve the influence of vehicle drag and rocket Isp. The thrust, Isp, and operating time of the rocket would determine the combination form ("T+RBCC" or "R+TBCC") of the rocket and TBCC. In this section, the study of the vehicle drag and the rocket Isp is conducted, and the combination form for the rocket and TBCC is also discussed.

Figure 11 shows the total weight consumption as a function of rocket thrust under different drag coefficients. To represent vehicles with different drag coefficients, AF_D is introduced in the present study and defined as the amplification factor of the drag coefficient of a specified vehicle to that of the baseline vehicle. Under an AF_D of 1.0 and 1.1, the TBCC engines could propel the vehicle to accelerate in the wide-speed range without a rocket. As the AF_D is increased, the TBCC thrust is unable to overcome the drag in the thrust pinch regions. Consequently, the rocket is introduced to augment the thrust. For each TBCC-powered vehicle, the thrust of the rocket has an optimal value to minimize the total weight consumption. With the increase in the AF_D, the minimum value and the optimal value of the relative thrust both increase. Although the introduction of rockets could propel the vehicle to accelerate in the wide-speed range, the total weight consumption in the process of acceleration increases with the increase in drag.

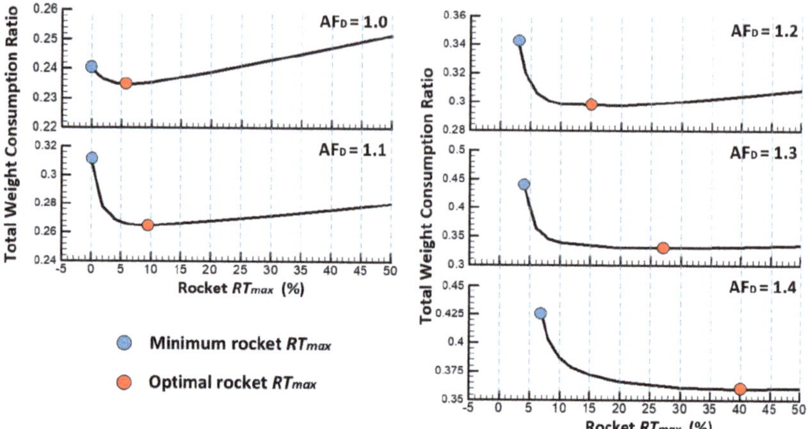

Figure 11. Effect of rocket RT_{max} on total weight consumption under different AF$_D$.

The optimal relative thrust of the rocket is presented in Figure 12 under various AF$_D$ and Isp. In the low-AF$_D$ cases (such as AF$_D$ = 1, 1.1), the optimal relative thrust is insensitive to the Isp. A possible reason is that the rocket works for a short time and the consumption is relatively small. Compared to the total fuel consumption, the fuel reduction caused by the increase in the rocket Isp could be negligible. In the high-AF$_D$ cases (such as AF$_D$ = 1.3, 1.4), the changes in the optimal relative thrust with Isp are obvious, because the operating time of the rocket is too long to ignore the impact of Isp. The higher the Isp, the less the reduction in the engine efficiency caused by the rocket, and the more the increase in vehicle efficiency. Thus, the optimal relative thrust increases to satisfy the requirement of higher vehicle efficiency. The results suggest that it is not necessary to pay much attention to the Isp of the rocket in the case of low drag while the Isp is important in the case of high drag.

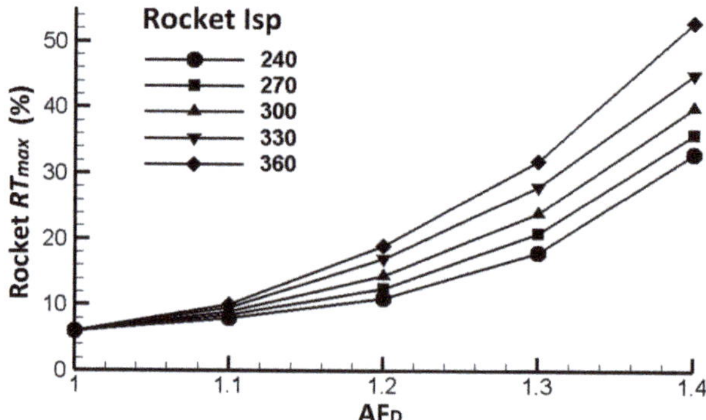

Figure 12. Optimal rocket RT_{max} under various AF$_D$ and rocket Isp.

Figure 13 shows the thrust profiles of the rocket under different AF$_D$ over the entire Mach range. To compare the thrust, the combined variable of AF$_D$ and rocket relative thrust RT is on the horizontal axis. It is to be noted that the RT is the product of the maximum relative thrust, RT_{max}, and the throttle, *Thr*. The distance from the solid line to the dotted line represents the relative thrust of the rocket. With the increase in AF$_D$, the operating range of the rocket expands in transonic and mode transition, and the value of the optimal thrust increases. This could be explained in that the expansion and deepening of the thrust gaps with the increase in the AF$_D$ call for a bigger thrust and a longer operation time of the rocket. In addition, under each AF$_D$, the rocket almost operates at the corresponding maximum thrust constantly in the operating range. It means the fixed-thrust design of the rocket could satisfy the thrust requirement.

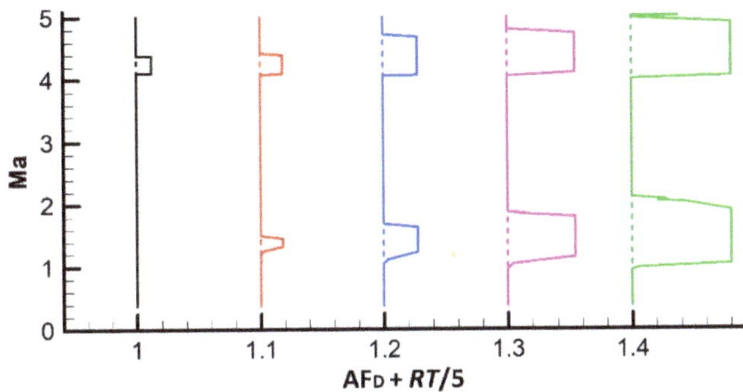

Figure 13. Rocket *RT* profiles at different AF$_D$ (Isp = 300 s).

Similarly, Figure 14 shows the thrust profiles of the rocket under different Isp over the entire Mach range. Under the AF$_D$ of 1.1 but different Isp, the maximum values of the thrust are approximately equal. The higher the Isp, the less the reduction in the engine efficiency caused by the rocket. Therefore, the operating ranges of the rocket are widened. In the case of Isp = 240 s, the rocket hardly works in the transonic, while in the case of Isp = 360 s, it operates in the range of Mach 1.18 to 1.56.

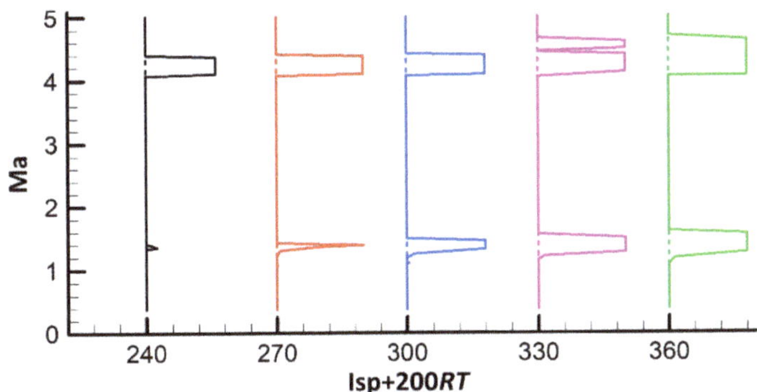

Figure 14. Rocket *RT* profiles under different rocket Isp (AF$_D$ = 1.1).

No matter what AF$_D$ and Isp are, the thrust is basically equal to the two values of 0 and maximum. This indicates that the thrust adjustment of the rocket is not necessary, and the fixed-thrust design could satisfy the thrust requirement and simplify the structure of the rocket.

Figure 15 shows the average vehicle efficiency, $\overline{\eta}_{veh}$, and the average engine efficiency, $\overline{\eta}_{engine}$, corresponding to different Isp and AF$_D$. With the increase in Isp, the $\overline{\eta}_{veh}$ increases due to the expansion of the rocket operating range, while the $\overline{\eta}_{engine}$ is roughly constant after including the effect of the improved engine efficiency of the rocket. When AF$_D$ = 1.4, as the Isp increases from 240 s to 360 s, the $\overline{\eta}_{veh}$ is improved from 0.368 to 0.453, an increase of 23.1%. By comparison, when AF$_D$ = 1.0, the Isp has little effect on the $\overline{\eta}_{engine}$ and the $\overline{\eta}_{veh}$, because the operating range of the rocket is very small. With the increase in AF$_D$, the $\overline{\eta}_{veh}$ decreases first and then increases. In low-AF$_D$ cases, the TBCC provides most of the thrust for acceleration, and the rocket works within a narrow range, serving as auxiliary power. The increase in drag decreases the $\overline{\eta}_{veh}$. By comparison, in high-AF$_D$ cases, the TBCC works independently in a narrow range. The rocket widens the operating range and increase the thrust of itself, and becomes a great part of the propulsion system. Due to the increase in rocket thrust, the $\overline{\eta}_{veh}$ increases.

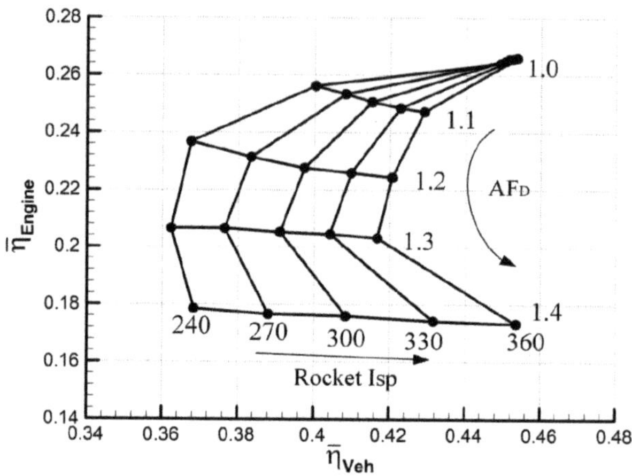

Figure 15. Effect of rocket Isp and vehicle drag on $\overline{\eta}_{engine}$ and $\overline{\eta}_{veh}$.

Figure 16 shows the average total efficiency, $\overline{\eta}_{tot}$, and the total energy consumption of drag, E_{Drag}, corresponding to different Isp and AF$_D$. With the increase in Isp, the $\overline{\eta}_{tot}$ increases, because the $\overline{\eta}_{veh}$ increases, and the $\overline{\eta}_{engine}$ remains almost unchanged. Meanwhile, the operating range of the rocket is widened and the E_{Drag} is reduced. When AF$_D$ = 1.4, as the Isp increases from 240 s to 360 s, the $\overline{\eta}_{tot}$ is improved from 0.066 to 0.079, namely an increase of 19.7%, and the E_{Drag} is decreased from 35.37 GJ to 24.71 GJ, namely a decrease of 30%.

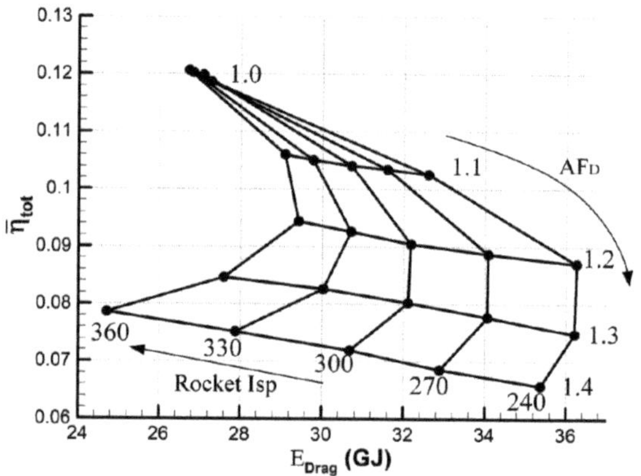

Figure 16. Effect of rocket Isp and vehicle drag on $\overline{\eta}_{tot}$ and E_{Drag}.

The increase in AF$_D$ decreases the $\overline{\eta}_{tot}$. Moreover, with the increase in AF$_D$, the E_{Drag} increases first and then decreases. The role played by the rocket accounts for this trend of E_{Drag}. When the rocket serves as an auxiliary power at low AF$_D$, the increase in AF$_D$ decreases the acceleration, resulting in an increase in the E_{Drag}. When the rocket serves as a great part of the propulsion system at high AF$_D$, the increase in AF_D widens the operating range of the rocket and increases the thrust of the rocket, which results in a decrease in the E_{Drag}.

Figure 17 presents the total weight consumption ratio and flight time under different Isp and AF_D. With the increase in Isp, the total weight consumption and the flight time are gradually reduced, because the $\overline{\eta}_{tot}$ increases and the E_{Drag} decreases. As the AF_D increases, the total weight consumption is increased.

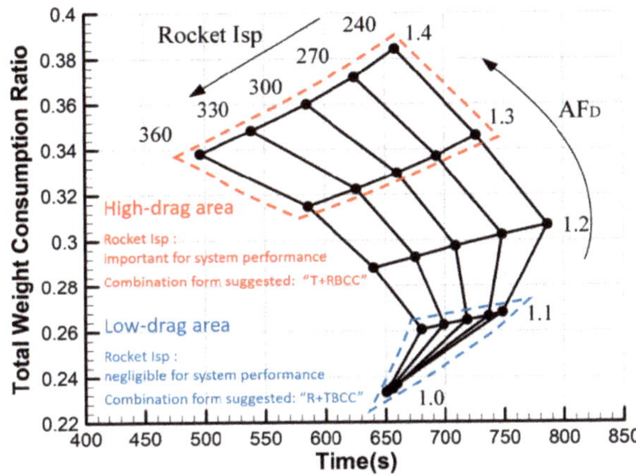

Figure 17. Total weight consumption ratio and flight time under different rocket Isp and AF_D.

The effect of Isp on weight consumption and flight time varies considerably under different AF_D. Therefore, this network map can be roughly divided into two areas: low-drag area and high-drag area. In the low-drag area, the rocket works for a short time and the optimal relative thrust is small. The consumption of the rocket is relatively small compared to the total fuel consumption. It is concluded that the improvement of the Isp of the small rocket is not necessary. In this case, the small rocket could serve as an additional booster, whose structure is simple. The combination of the rocket and TBCC engine is similar to the "R+TBCC". By contrast, in the high-drag area, the operating time of the rocket is long and the optimal relative thrust is large. The Isp has a significant impact on the flight time and the total fuel consumption. It is necessary to pay attention to the Isp of the big rocket. Introducing a big rocket as a booster might bring a large base drag. Instead, the rocket should be integrated into the TBCC engine, such as an RBCC or ERJs, which could deliver augmented thrust at an Isp performance of up to twice or triple that of a rocket through proper design. The combination of the rocket and TBCC engine is similar to the "T+RBCC".

According to the parameter study of the vehicle drag and rocket Isp, two design recommendations for rocket-augmented TBCC engines corresponding to the low-drag area and the high-drag area are as follows. In the low-drag area, the rocket works for a short time and its optimal relative thrust is small, so optimizing the Isp of the small rocket is not necessary. The TBCC combined with a booster rocket could be a more simple and suitable scheme. By comparison, in the high-drag area, the operating time of the rocket is long and the optimal relative thrust is large. The Isp has a significant impact on the flight time and the total fuel consumption. Therefore, combining an RBCC (or ERJs) and a turbine could produce higher economic performance.

6. Conclusions

In this study, an integrated analysis method for rocket-augmented TBCC-powered vehicles was proposed based on the trajectory optimization method of the Gauss pseudospectral. The trajectories of the baseline vehicle with and without a rocket were analyzed from the aspect of efficiency and energy.

The combination form for rockets and TBCC was discussed with the parameter study of vehicle drag and rocket Isp. Accordingly, the following conclusions were obtained.

(1) Introducing an appropriate rocket to assist the TBCC-powered vehicle could reduce the total energy consumption of drag and increase the vehicle efficiency in the transonic and the mode transition. It results in an increase in total efficiency in spite of the reduction in engine efficiency. Therefore, introducing the rocket as the auxiliary power is not only a practical solution to enable flight over a wide-speed range when the TBCC is incapable but also an economical scheme when the TBCC meets the requirements of thrust.

(2) For a specified TBCC-powered vehicle, there is an optimal thrust of the rocket to maximize the economic performance. That is to say, the rocket with an overly small thrust could not improve the vehicle efficiency sufficiently, while a rocket with an overly large thrust might lead to excessive drag. In addition, the rocket almost stays at the maximum thrust constantly when it works. It means the fixed-thrust design of the rocket could satisfy the requirement.

(3) When the vehicle drag is low, the rocket works for a short time and its optimal relative thrust is small, thus optimizing the Isp of the small rocket is not necessary. The TBCC combined with a booster rocket will be a very simple and suitable scheme. When the vehicle drag is high, the operating time of the rocket is long and the optimal relative thrust is large. The Isp has a significant impact on the flight time and the total fuel consumption. Therefore, the combination form for an RBCC (or ERJs) and the turbine will be more appropriate to obtain higher economic performance.

Author Contributions: Conceptualization, F.G. and J.Z.; methodology, F.G. and J.Z.; software, F.G.; validation, F.G. and W.L.; formal analysis, F.G. and J.Z.; investigation, F.G.; data curation, F.G.; writing—original draft preparation, F.G.; writing—review and editing, F.G., W.L. and J.Z.; visualization, F.G.; supervision, F.X.; project administration, J.Z. and Y.Y.; resources, Y.Y.; funding acquisition, Y.Y. All authors have read and agreed to the published version of the manuscript.

Funding: This study was funded by the National Natural Science Foundation of China (No. 51906208 and No. 91941103), the Equipment Exploratory Research Fund (No. 61402060301), the Aeronautics Power Foundation (No. 6141B090325) and the Weapon Exploratory Research Fund (No. 6141B010266).

Acknowledgments: The team members of School of Aerospace Engineering of Xiamen University are gratefully acknowledged. In addition, Feng Guo wants to thank the care and support from Shengnan Dai over the passed years.

Conflicts of Interest: The authors declare no conflicts of interest.

References

1. Hueter, U.; McClinton, C. NASA's Advanced Space Transportation Hypersonic Program. In Proceedings of the 11th AIAA/AAAF International Conference Space Planes and Hypersonics Systems and Technologies Conference, Orleans, France, 29 September–4 October 2002; AIAA: Reston, VA, USA, 2002.

2. Liu, J.; Yuan, H.; Hua, Z.; Chen, W.; Ge, N. Experimental and numerical investigation of smooth turbine-based combined-cycle inlet mode transition. *Aerosp. Sci. Technol.* **2017**, *60*, 124–130. [CrossRef]

3. Bulman, M.; Siebenhaar, A. Combined Cycle Propulsion: Aerojet Innovations for Practical Hypersonic Vehicles. In Proceedings of the 17th AIAA International Space Planes and Hypersonic Systems and Technologies Conference, San Francisco, CA, USA, 11–14 April 2011.

4. Siebenhaar, A.; Bogar, T. Integration and Vehicle Performance Assessment of the Aerojet "TriJet" Combined-Cycle Engine. In Proceedings of the 16th AIAA/DLR/DGLR International Space Planes & Hypersonic Systems & Technologies Conference, Bremen, Germany, 19–22 October 2009.

5. Wei, B.; Ling, W.; Luo, F.; Gang, Q. Propulsion Performance Research and Status of TRRE Engine Experiment. In Proceedings of the 21st AIAA International Space Planes and Hypersonics Technologies Conference, Xiamen, China, 6–9 March 2017.

6. Yang, H.; Ma, J.; Man, Y.; Zhu, S.M.; Ling, W.; Cao, X.B. Numerical Simulation of Variable-Geometry Inlet for TRRE Combined Cycle Engine. In Proceedings of the 21st AIAA International Space Planes and Hypersonics Technologies Conference, Xiamen, China, 6–9 March 2017.

7. Guo, F.; Gui, F.; You, Y.C.; Zhu, J.F.; Zhu, C.X. Experimental Study of TBCC Engine Performance in Low Speed Wind Tunnel. *J. Propuls. Technol.* **2019**, *40*, 2436–2443.

8. Pastrone, D.; Sentinella, M.R. Multi-objective optimization of rocket-based combined-cycle engine performance using a hybrid evolutionary algorithm. *J. Propuls. Power* **2009**, *25*, 1140–1145. [CrossRef]

9. Dijkstra, F.; Maree, A.G.M.; Caporicci, M.; Immich, H. Experimental investigation of the thrust enhancement potential of ejector rockets. In Proceedings of the 33rd Joint Propulsion Conference and Exhibit, Seattle, WA, USA, 6–9 July 1997.

10. Etele, J.; Parent, B.; Sislian, J.P. Analysis of Increased Compression through Area Constriction on Ejector-Rocket Performance. *J. Spacecr. Rockets* **2007**, *44*, 355–364. [CrossRef]

11. Serre, L.; Defoort, S. LAPCAT II: Towards a Mach 8 civil aircraft concept, using advanced Rocket/Dual-mode ramjet propulsion system. In Proceedings of the 16th AIAA/DLR/DGLR International Space Planes & Hypersonic Systems & Technologies Conference, Bremen, Germany, 19–22 October 2009.

12. Li, Y.; Jiang, G.T.; Chu, X.Y.; Wang, J. Research on TBCC Engine Size Selection and Ascent Strategy of Combined-Cycle Aircraft. *J. Astronaut.* **2018**, *39*, 17–26. [CrossRef]

13. Berend, N.; Talbot, C. Overview of some optimal control methods adapted to expendable and reusable launch vehicle trajectories. *Aerosp. Sci. Technol.* **2006**, *10*, 222–232. [CrossRef]

14. Stancil, R.T. A New Approach to Steepest-Ascent Trajectory Optimization. *AIAA J.* **1964**, *2*, 1365–1370. [CrossRef]

15. Li, Z.H.; Hu, C.; Ding, C.B.; Liu, G.; He, B. Stochastic gradient particle swarm optimization based entry trajectory rapid planning for hypersonic glide vehicles. *Aerosp. Sci. Technol.* **2018**, *76*, 176–186. [CrossRef]

16. David, A.B.; Geoffrey, T.H.; Tom, P.T.; Anil, V.R. Direct trajectory optimization and costate estimation via an orthogonal collocation method. *J. Guid. Control Dyn.* **2006**, *29*, 1435–1440.

17. Ross, I.M. A Roadmap for Optimal Control: The Right Way to Commute. *Ann. N. Y. Acad. Sci.* **2006**, *1065*, 210–231. [CrossRef] [PubMed]

18. Yang, S.; Cui, T.; Hao, X.Y.; Yu, D.R. Trajectory optimization for a ramjet-powered vehicle in ascent phase via the Gauss pseudospectral method. *Aerosp. Sci. Technol.* **2017**, *67*, 88–95. [CrossRef]

19. Clough, J.A.; Lewis, M.J. Comparison of Turbine-Based Combined-Cycle Engine Flowpaths. In Proceedings of the AIAA International Space Planes & Hypersonic Systems & Technologies, Norfolk, VA, USA, 15–19 December 2003.

20. Huang, W.; Yan, L.; Tan, J.G. Survey on the mode transition technique in combined cycle propulsion systems. *Aerosp. Sci. Technol.* **2014**, *39*, 685–691. [CrossRef]

21. Heiser, W.H.; Pratt, D.T. *Hypersonic Airbreathing Propulsion*, 5th ed.; American Institute of Aeronautics and Astronautics, Inc.: Washington, DC, USA, 1994.

22. Brock, M.A. Performance Study of Two-Stage-To-Orbit Reusable Launch Vehicle Propulsion Alternatives. Air Force Institute of Technology. Master's Thesis, AIR Force Institute of Technology, Wright-Patterson AFB, OH, USA, 2004.

23. Bryson, J.A.E.; Desai, M.N.; Hoffman, W.C. Energy-state approximation in performance optimization of supersonic aircraft. *J. Aircr.* **1969**, *6*, 481–488. [CrossRef]

24. Parker, J.T.; Bolender, M.A.; Doman, D.B. Control-oriented modeling of an air-breathing hypersonic vehicle. *J. Guid. Control Dyn.* **2007**, *30*, 856–869. [CrossRef]

25. Zheng, J.L.; Chang, J.T.; Yang, S.B.; Hao, X.Y.; Yu, D.R. Trajectory optimization for a TBCC-powered supersonic vehicle with transition thrust pinch. *Aerosp. Sci. Technol.* **2019**, *84*, 214–222. [CrossRef]

26. Patterson, M.A.; Rao, A.V. GPOPS-II: A MATLAB Software for Solving Multiple-Phase Optimal Control Problems Using hp-Adaptive Gaussian Quadrature Collocation Methods and Sparse Nonlinear Programming. *ACM Trans. Math. Softw.* **2014**, *41*, 1–37. [CrossRef]

27. Darby, C.L.; Hager, W.W.; Rao, A.V. An hp-adaptive pseudospectral method for solving optimal control problems. *Optim. Control Appl. Methods* **2011**, *32*, 476–502. [CrossRef]

Review

A Review on Liquefied Natural Gas as Fuels for Dual Fuel Engines: Opportunities, Challenges and Responses

Md Arman Arefin [1], Md Nurun Nabi [2,*], Md Washim Akram [3], Mohammad Towhidul Islam [1] and Md Wahid Chowdhury [1]

[1] Department of Mechanical Engineering, Rajshahi University of Engineering & Technology, Rajshahi 6204, Bangladesh; arefinarman@gmail.com (M.A.A.); towhid1396@gmail.com (M.T.I.); devmario30@gmail.com (M.W.C.)

[2] School of Engineering and Technology, Central Queensland University, Perth, WA 6000, Australia

[3] Department of Mechanical Engineering, Bangladesh Army University of Science and Technology, Saidpur 5310, Bangladesh; washim@baust.edu.bd

* Correspondence: m.nabi@cqu.edu.au

Received: 22 October 2020; Accepted: 20 November 2020; Published: 23 November 2020

Abstract: Climate change and severe emission regulations in many countries demand fuel and engine researchers to explore sustainable fuels for internal combustion engines. Natural gas could be a source of sustainable fuels, which can be produced from renewable sources. This article presents a complete overview of the liquefied natural gas (LNG) as a potential fuel for diesel engines. An interesting finding from this review is that engine modification and proper utilization of LNG significantly improve system efficiency and reduce greenhouse gas (GHG) emissions, which is extremely helpful to sustainable development. Moreover, some major recent researches are also analyzed to find out drawbacks, advancement and future research potential of the technology. One of the major challenges of LNG is its higher flammability that causes different fatal hazards and when using in dual-fuel engine causes knock. Though researchers have been successful to find out some ways to overcome some challenges, further research is necessary to reduce the hazards and make the fuel more effective and environment-friendly when using as a fuel for a diesel engine.

Keywords: liquefied natural gas; diesel engine; greenhouse gas emissions; sustainable development

---

1. Introduction

Recently, it is projected that global energy demand has been increasing day by day [1,2]. The increased amount of energy demand produces a large amount of greenhouse gases (GHGs) by the burning of fossil fuels, which ultimately causes global warming. Currently, in the industrial and transportation sectors, diesel is mainly used as fossil fuel. The consumption of diesel fuel is increasing day by day unexpectedly because of the dramatic increase in vehicles, mostly in Asian countries like Bangladesh, Korea, India and China [3]. Currently, researchers all over the world are concerned about the way of mitigating this large amount of energy demand and at the same time, carbon dioxide (CO_2) emission reduction, which is one of the major components of GHGs [4,5]. In this regards, other sources of fuel may be a feasible alternative to conventional fuels. Natural gas would be a promising alternative fuel source in the transportation sector because of some of its remarkable advantages. Though natural gas is also derived from fissile resources, it can be converted from renewable sources, i.e., it can be produced through the biomass conversion process (biomethane, which is also known as biogas, is a pipeline-quality gas made from organic matter), attractive cost, better combustion efficiency and greenhouse gas reduction, which are the significant advantages as alternative fuel [3].

Besides, the interest in alternative fuels is rapidly growing because of the energy security concern all over the world. Among the candidates of alternative fuel, biofuels, liquefied petroleum gas (LPG) and liquefied natural gas (LNG) are the potential ones. Nevertheless, the economic aspects and availability make LPG and LNG more realistic solutions compared to biofuels. Additionally, natural gas can be considered as one of the solutions to control engine emissions at present as compared to traditional fuels. In a homogenous charge compression ignition engine (HCCI), the after-treatment technique combined with natural gas present an outstanding potential to meet strict requirements for reducing the emissions [6]. In the transportation sector, both forms of natural gas, i.e., liquefied natural gas (LNG) and compressed natural gas (CNG) can be used as an alternative fuel. However, CNG has already gained popularity in the automobile sector due to its lower cost.

LNG is mainly used for electricity production and transportation. The main advantages offered by LNG are higher safety, easier transportation and storage capacity compared to CNG [3]. LNG is cleaner than coal and oil, therefore, it has got a plethora of recognition in the global market [7]. Since LNG is clean, it is a promising alternative to diesel vehicles and is capable of compensating some of the severe drawbacks of natural gas vehicles; for instance, LNG fuelled trucks have a higher range (up to 700–1000 km) due to higher energy density [8]. However, while considering LNG as an alternative, the economic viability should also be taken into consideration. The cost of an LNG fuel tank and the engine is approximately twice as high as a diesel engine and tank [9]. Consequently, during this decade, the main contributors to LNG supply growth will be Australia and USA, though 18 other countries have already joined and others have a short and long term goal to join the industry [10]. Besides, among various alternatives of natural gas, LNG is the most preferred mode for long distance transportation as because of its liquefaction attribute, its volume reduces by a factor of 600 [11]. Compared to piped natural gas (PNG), the annual growth of LNG trade is two times higher, which accounts for 10% and 31% of global natural gas consumption and trade respectively [8]. According to recent study reports, from 2005 to 2015, Europe, Asia and Oceania were the primary recipients of LNG imports consisted of 90% of global imports [12].

1.1. Economics and Life Cycle of LNG

Figure 1 shows the worldwide LNG price as of June 2019. South Korea and China had the highest landed price of LNG in the world. The price received at the regasification plant is referred to as the landed price. Netback price is taken into consideration for the determination of these prices, which is based on the effective price for a seller and producer at a definite location.

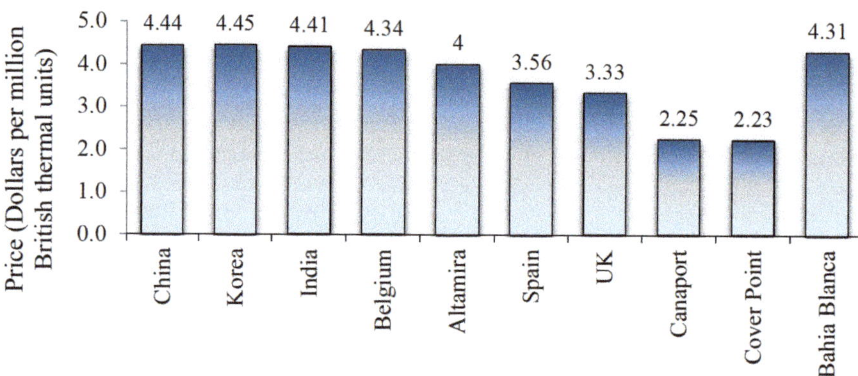

Figure 1. Worldwide landed price of liquefied natural gas (LNG) [13].

It is also required to make an economic calculation of the LNG liquefaction plant because fuel imposes the second highest operational cost after labor. There are some reasons behind the cost of

LNG not being as straight forward as diesel, which are: (1) distribution costs are not included in LNG fuel price, which is dissimilar to diesel fuel price; (2) generally state tax rates for LNG is different from other fuels, and it also varies state-wise; (3) federal tax rates for LNG fuel are also different and (4) energy content of different fuels are different [14]. Table 1 shows the national average cost for LNG and diesel fuel. Equivalent fuel cost is calculated in dollars per diesel equivalent liter by using the energy content of the fuel. Diesel equivalent liter refers to the amount of energy, which is equivalent to one-liter diesel fuel. Since the distribution cost is not included in LNG fuel price, thus, it is more competitive. Based on the distribution costs and state taxes, the costs of LNG fuel will vary radically by location to location. According to the Taxpayer Relief Act of 1997, based on the energy content of transportation fuel it has been taxed, such as the federal tax rate on LNG is changed from $0.0503 (unit conversion) per liter to USD 0.0317 (unit conversion) per liter [14].

Table 1. Average LNG as well as diesel fuel cost worldwide (unit conversion) [14].

Fuel Type	Fuel Cost ($/L)	Distribution Cost ($/L)	State Tax ($/L)	Federal Tax ($/L)	Total Fuel Cost ($/L)	Equivalent Fuel Cost ($/Diesel Equivalent Liter)
Diesel	0.1717		0.0634	0.0555	0.2906	0.2906
LNG	0.0925	0.0264	0.0476	0.0660	0.2325	0.3936

Note: Permission granted from National Renewable Energy Laboratory.

Figure 2 depicts a schematic diagram of the LNG life cycle. With the help of the pipeline extracted raw natural gas is sent to the plant for liquefaction, where LNG is obtained as a byproduct. Mainly, the ship is used to carry LNG to the regasification terminal. Meanwhile, a small amount of liquid is drawn off and carried by small tanker trucks to the service station for fuelling LNG trucks. On the other hand, in each transport and storage, phase boil-off gas is produced, which can be fully recovered for the purpose of using as a gaseous fuel [15].

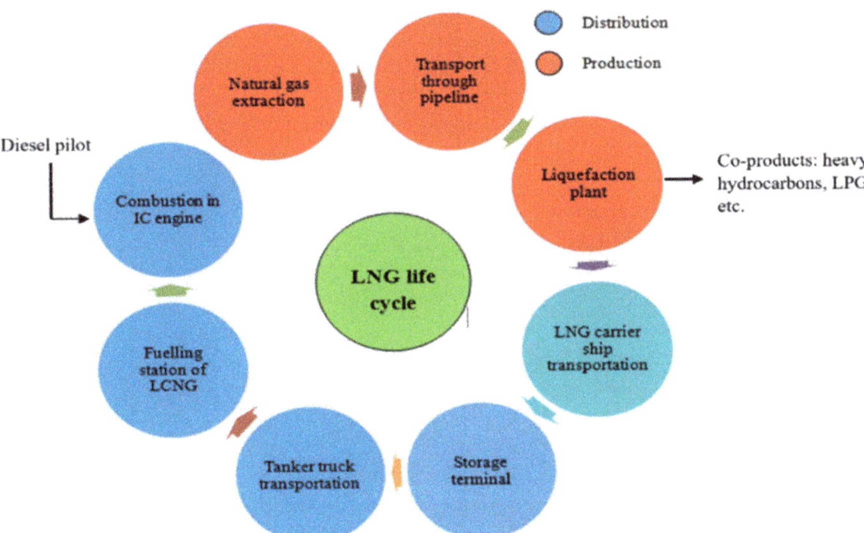

Figure 2. LNG life cycle (adapted from [15]). Note: Permission granted from the publisher (Elsevier-Applied Energy).

Although natural gas is benign and available in nature, it is hardly used in a large diesel engine and still limited to a small engine i.e., spark ignition (SI) engine. Natural gas is mainly used in heavy diesel operated transportation fleets [3]. It is reported that fuelling natural gas with diesel fuel as compared to fuelling diesel only causes a less lean mixture and a lower the efficiency. This is because when natural gas is fuelled it tends to mix with air and could move into the cylinder. Hence, this induced less air leads to an increase in equivalence ratio. Again, when the load is decreased, it causes lesser inducing of natural gas, which results in reducing the efficiency in fixed load speed condition and, thus, more natural gas is needed to be aspirated [16]. Besides, knocking is one of the severe problems of dual-fuel engines [17–19]. According to previous study reports, though advancing the injection timing in the dual-fuel engine reduces emissions [20], may increase the tendency of noise due to fast pressure rise [21,22]. Thus, natural gas is always used as minor fuel in most of the dual-fuel engine operations [3]. One of the key initiatives to utilize LNG as a fuel in a diesel engine is to alleviate the knocking problem and take adequate precautions when knocking resistance is low [23]. In terms of emissions, LNG provides far better results compared to conventional fuels. For instance, an experiment in the Netherlands showed that for trucks GHG emission will be reduced by 10–15% if LNG is used [24]. With the increase in power, the smoke density of diesel increases sharply while LNG-duel fuel increases slightly. Under high loads, the smoke density of LNG duel fuel decreases by a huge amount when compared with diesel [25].

From the above discussion, it can be said that with the increasing demand for natural gas vehicles, the demand for higher performance, environment-friendly and more efficient engines have also been increased. The perspective of this review is to bring attention to the current position of LNG and the use of it in the diesel engine. Engine performances and emission characteristics by using LNG are also discussed in this study. The consciousness of using environmentally viable and economical energy source is attracting researchers and environmental scientists to concentrate on LNG as fuels to fulfill both energy crisis and to control environmental abnormalities.

1.2. Properties of LNG

LNG is a mixture of gases, and its liquefaction is done by reducing the temperature below the boiling point. The amount of methane in LNG is about 87–99 mole%, and the remaining portion is ethane, propane and other heavier hydrocarbons depending on different LNG sources [26,27]. For instance, the LNG imported from Belgium contains 90% (by mass) methane and 10% (by mass) ethane [28]. The lower calorific value of LNG is 21 MJ/L, and the higher calorific value is 24 MJ/L at −164 °C [29]. To produce LNG, natural gas is refrigerated at −162 °C at atmospheric pressure; thus, LNG is known as a cryogenic liquid [30,31]. During liquefaction of natural gas, the primary component, i.e., methane, is cooled below its boiling point. At the same time, the concentrations of oxygen, carbon dioxide, water, hydrocarbons and some sulphur compounds are either removed or reduced in some small extent [32]. Normally, LNG is stored and handled below 1.586 MPa (unit conversion) pressure in extremely cold temperature in the tank [33]. The density of LNG is between 410 and 500 kg/m^3 [29], hence, it is lighter than water [28] and will float if it spilled on water. Per unit volume combustion heat production of LNG is much higher compared to natural gas. At atmospheric conditions, to produce equal energy, natural gas requires 600 times larger volume compared to LNG [28]. Besides, both LNG and its vapor are not explosive when exposed to the unconfined environment [32]. Moreover, LNG is more efficient and feasible for transportation compared to pipeline gases [7,29,34]. Well purified and condensed LNG can be easily transported over the sea [35]. While transporting, to handle the low temperature of LNG, specifically designed double-hulled ships are used [32].

LNG is a non-toxic, non-corrosive, colorless, odorless, safe and clean form of natural gas [7,36]. LNG is non-flammable; therefore, the liquid itself will not burn. However, the vapor of LNG is highly flammable with air, which causes flash fire. The flashpoint of LNG is −187.8 °C, but the autoignition temperature is 537 °C [37]. The specific gravity of LNG is 0.45 and 0.6 of gas [37] and 2.7488 liters of LNG has 100% of the energy of 3.7854 liters (unit conversion) of diesel [38]. The burning speed of LNG is 0.38 m/s in the stoichiometric mixture. This burning speed is comparatively lower for using only LNG in a diesel engine [37]. Due to the requirement of thermal shield and the lower density of

LNG compared to heavy fuel oil (HFO), the fuel tank required for LNG is 2.5–3 times bigger than HFO tank [39]. During combustion, LNG has almost no SO_2 and particulate matter emission [40]. Furthermore, the life cycle CO_2 emissions of LNG are 18% less than its counterpart gasoline vehicle model [41]. These advantages enable LNG as a potential fuel for the transportation sector. Depending on the liquefaction process and the plant where LNG is produced, the chemical components of LNG vary slightly, i.e., Europe, America and Asia, shown in Table 2 [42]. During designing LNG power plant, the regasification and liquefaction processes are controlled by the phase behavior and thermodynamic properties of LNG. Mokhatab et al. [26] have summarized the multiphase equilibrium of LNG using numerical methods.

Table 2. Chemical compositions of LNG imported from various countries all over the world [42].

Terminal	Methane	Ethane	Propane	Butane	Nitrogen
Abu Dhabi	87.07	11.41	1.27	0.14	0.11
Alaska	99.8	0.10	NA	NA	NA
Algeria	91.40	7.87	0.44	0.00	0.28
Australia	87.82	8.30	2.98	0.88	0.01
Brunei	89.40	6.30	2.80	1.30	0.00
Indonesia	90.60	6.00	2.48	0.82	0.09
Malaysia	91.15	4.28	2.87	1.36	0.32
Oman	87.66	9.72	2.04	0.69	0.00
Qatar	89.87	6.65	2.30	0.98	0.19
Trinidad	92.26	6.39	0.91	0.43	0.00
Nigeria	91.60	4.60	2.40	1.30	0.10

Note: Reprinted with permission from the publisher (Elsevier-Renewable and Sustainable Energy Reviews).

LNG is not explosive; therefore, to be ignited, first, it must be vaporized and then mixed with air at a proper portion [43]. Nevertheless, in the LNG transportation sector, there are some possibilities of accident, for example, the leakage of LNG into ground and sea, and rollover of the LNG tank [35]. However, LNG vapor only burns when it is mixed in a concentration of 5%–15% with air [7,32]. All these factors make LNG a safe alternative fuel for the transportation sector.

LNG makes two distinct types of gases, depending on the extraction procedure. They are natural boil-off gas and forced boil-off gas. From the top of the fuel tank, natural boil-off gas is collected, which contains a higher amount of methane and a lower amount of nitrogen; thus, its knocking resistance is very high. Investigation predicts the value of methane number (MN) is around 100 and the lower calorific value between 33 and 35 MJ/nm^3 [44].

On the other hand, from the down in the tank, forced boil-off gas is extracted and evaporated separately. Forced boil-off gas contains the mixture of all hydrocarbons of the liquid. The knocking resistance of the derived gas varies from load to load and even from origin to origin with the methane number ranging between 70 and 80. The calorific value of forced boil-off gas is about 38–39 MJ/nm^3, which is quite high compared to the natural boil-off gas [44]. Besides, the forced boil-off gas is quite stable than the natural boil-off gas. Thus, for general shipping forced boil-off gas is becoming very popular [44]. After LNG being heated by ambient air, boil-off gas is generated from the tank at the lowest boiling point of the constituents. Thus, it is very difficult to keep the LNG compositions unchanged for a long time once it started to boil-off. Since heavy trucks are operated in the same routes thus, LNG is more suitable as the fuel of those engines. Furthermore, the cruising distance will be tripled if LNG is used as a fuel in the vehicle instead of CNG because the energy density of CNG is three times lower than LNG [45].

Table 3 depicts the comparison of the physical properties of LNG, CNG, diesel and gasoline. With the change of chemical compositions, the physical properties of LNG vary like diesel and gasoline. In some cases, like the pump octane number and lower heating value, CNG and LNG show somewhat similar results. It is important to understand the cryogenic property of LNG during using it as a vehicle fuel. The phenomenon "weathering" or "enrichment" arises because natural gas is a chemical mixture of ethane, nitrogen and methane. LNG contents vary in the percentage of methane as well as other hydrocarbons. Generally, methane content may vary from 92% to 99% [14]. Besides, other hydrocarbon contents in natural gas are ethane 1%–6%, butane 0% to 2%, propane 1%–4% and other compounds [14]. The boiling point of each of the chemical compounds is methane, ethane, propane and butane is −162, −88, −42 and −0.5 °C respectively [45]. This is the reason why the chemical compounds in the liquid vaporize at its unique boiling point. Since the boiling point of LNG components are different and it vaporizes at its unique boiling point, the concentration of heavier hydrocarbons of LNG is increased. Therefore, premature ignition occurs, which tends to "knock" due to the higher concentrations of heavier hydrocarbons. To relieve this uncontrolled knocking tendency, it is required to use LNG before it becomes weathered. To distinguish this potential difficulty, during the manufacture of LNG, it should be ensured that the amount of methane is 99.4% or higher [14]. The higher amount of methane reduces the detrimental constituents thus weathering cannot create harmful fuel mixtures in LNG [14].

Table 3. Comparison of different properties of LNG with CNG, diesel and gasoline.

Property	LNG	CNG	Diesel	Gasoline
Phase	Cryogenic liquid [14]	Gas [45]	Liquid [45]	Liquid [46]
Fuel Material	Underground reserves and renewable biogas [38]	Underground reserves and renewable biogas [38]	Crude Oil [38]	Crude Oil [38]
Composition	CH_4 (vol.) 99.80% [3] C_2H_6 (vol.) 0.10% [3] N_2 (vol.) 0.10% [3]	CH_4 (mole) 84.5% [47] C_2H_6 (mole) 7.70% [47] C_3H_8 (mole) 2.40% [47] C_4H_{10} (mole) 0.58 [47] C_5H_{12} (mole) 0.37 [47]	Typically, alkanes, polyaromatics, cycloalkanes or naphthenes [48]	Alkanes (vol.) 4–8% [46] Alkenes (vol.) 2–5% [46] Iso-alkanes (vol.) 25–40% [46] Cycloalkanes (vol.) 3–7% [46] Cycloalkenes (vol.) 1–4% [46] Total Aromatics (vol.) 20–50% [46]
Density (kg/m^3)	435 at 20 MPa [7]	175 at 20 MPa [7]	850.4 at 15 °C [3]	742.92 (unit conversion) [14]
Pump Octane Number	120+ [38]	120+ [38]		84–93 [38]
Lower heating value (kJ/kg)	49,244 [3]	46,892.16 (unit conversion) [38]	43,400 [3]	44,500 [46]
Cetane Number	NA [38]	NA [38]	40–55 [38]	NA [38]
Toxic	No [7]	No [7]	Yes [7]	Yes [7]
Health hazards	None [7]		None [7]	Eye irritant [7]

2. LNG Fuel System

There are mainly three types of the engine that burns LNG as combustion fuel in the marine industry. First one is the "spark ignited" engine, which uses only LNG as fuel with major requirements, simplicity and good overall performance at the lowest total emissions. This engine is generally used in the marine industry for short-distance ferries [44]. The second one has the dual-fuel capability and mainly used for land power plant because of higher specific power production capacity. The third one is the "direct gas injection diesel engine", which was firstly used in the offshore industry because of its high fuel flexibility and power density [44]. The updated concept of this type of engine in the marine industry has made it too flexible but the use of this type is still limited. The multifuel capability, flexibility in fuel mixing and high-pressure gas injection (300–350 bar) made this engine very promising. This type of engine is unique in terms of having no distinct requirements for stabilization of self-ignition of fuel gas. In addition, its operating principle for diesel provides a complete combustion of the gas fuel, though the NOx emission occurred higher than other gas engine types.

2.1. Descriptions of the Additional Components

Generally, diesel fuel containers are a low-pressure tank and kept uninsulated. However, LNG must be kept in a very well insulated and pressurized tank to minimize vapor loss and weathering because of its cryogenic nature [49]. To utilize LNG in a vehicle, a fuel control system is required, which consists of a pressure management system, vaporizer, fill and vent connections and a cryogenic tank. These devices are pre-plumbed and often integral to the tank. Tank pressure depends on fuel temperature and corresponding vapor pressure. Primarily, the saturated vapor pressure is lower than the fill pressure pumped. However, for a given fuel temperature, the saturation pressure will rapidly drop, and condensation of vapor takes place into the liquid. Most of the systems entirely depend on fuel saturation pressure and temperature, but there are some special devices with systems, which can add heat during lowering the tank pressure [50]. Between the inner and outer shell of the fuel tank, there is a perlite insulation to reduce the heat leakages in the vacuum and to maintain the heat in-between the tanks [51].

All the fuel lines, fittings and valves between LNG fuel tank and the vaporizer must be capable of providing cryogenic services. The connecting lines from the vaporizer to the engine are required only to carry the fuel vapor to the engine inlet. It is recommended to use corrosion-resistant fittings to avoid contamination. Total pressure drops to deliver fuel between the fuel tank and engine inlet must be adequate. The pressure drop testing is necessary to measure the performance and isolate the components if excessive system pressure drop occurs and then, system modification is also necessary [50]. Among various types of vaporizers, the intermediate fluid vaporizer (IFV) shows several advantages over others. For designing the IFV, the heat source must be determined considering the climatic conditions and environmental issues [52].

Besides the cryogenic tank, liquid level and pressure transmitters, venting system, cryogenic filling connector and pressure relief valve are other important components. The main feature of the pressure relief valve is to control the boil-off of the tank. To take the liquid from the tank and then pass it through a pressure regulator and vaporizer to the engine feeding system and injector a withdrawal system is used, which is associated with the cryogenic tank. However, there are some modern engines that are designed to directly inject liquid natural gas in the combustion chamber. Due to having a boil-off valve with a cryogenic container, LNG vehicle should be operated continuously without long parking [53].

2.2. Possible Outcome

From a long time, natural gas has been used as fuel for private cars. Heavy vehicles such as heavy trucks and buses are also using natural gas recently. Pressure vessel stores natural gas at very high pressure of about 20 MPa. In the transport sectors, natural gas shows numerous benefits, especially where heavy fuels are utilized [28]. Almost 25% CO_2 [54], 85% NOx [28], 98% SOx [28] and 90% particulate matters [28] are reduced during combustion of the engine when natural gas is used in place of diesel fuel. Shortly, heavy fuels containing sulphur will be severely restricted; thus, Western Europe has a considerable interest to use LNG for ship propulsion [28]. Worldwide share of ship emissions to total emissions is increasing day by day. Ship emission produces SO_2, CO_2 and NO_X, which contributes 4–8%, 2–4% and 10–20% of global emissions, respectively [55]. In the gas-burning mode, LNG vessels result in the elimination of all SO_2 emission and the production of NO_X, CO_2 and particulate matter are less compared to a vessel driven by marine diesel [55]. Generally, the greenhouse gas emission of a diesel truck is higher than the LNG truck in the wheel to wheel (WTW) analysis. Figure 3 shows a particularity of source of origin of tank to wheel (TTW) emission factors, where DENA extricated the information from German Consultancy-Ludwig Bölkow Systemtechnik (LBST) and the rest of the sources are from North American experiments [34]. From Figure 4, it can be seen that the total WTW emissions of all gas heavy ground vehicles (HGVs) are significantly lower than diesel HGVs and LNG shows way better results in terms of emissions than CNG and diesel. Besides, the reduction of emission, LNG provides other competitive advantages in night-time services through inner-city by reducing the noise level when using LNG Otto-cycle engines [56].

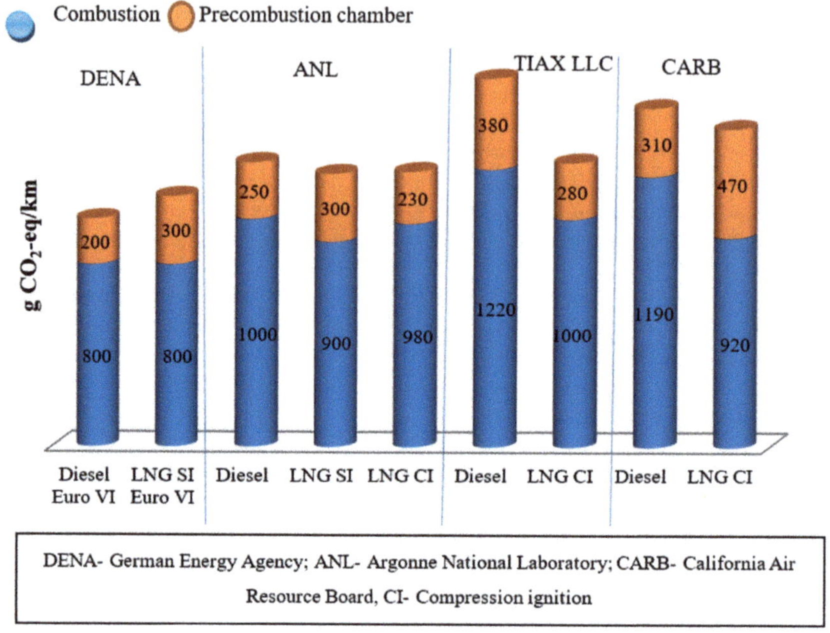

Figure 3. Life cycle greenhouse gas comparison of diesel and LNG heavy-duty vehicle (adapted from [40]). Note: Permission granted from the publisher (Elsevier-Renewable and Sustainable Energy Reviews).

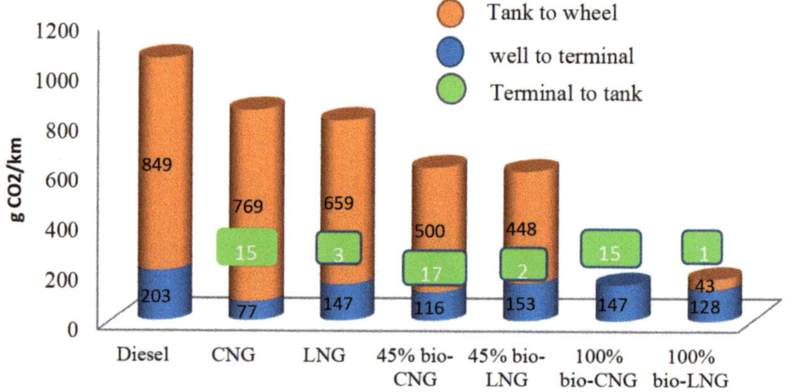

Figure 4. Wheel to Wheel emissions of heavy goods vehicles, adapted from [57].

For many years, the price of LNG depends on the HFO price though often LNG is cheaper. By considering the lower heating value and price of LNG, it can be said that LNG cost is about 60% of HFO [37]. Due to the savings of the reliquefaction process, the cost of produced boil-off gas is decreasing on gas carriers. Due to introducing shale gas in the US market, the natural gas price, including LNG, decreased in 2009 and 2010. Therefore, LNG has become a strong competitive of HFO. The price of HFOIF-0380 in the middle of 2008 was over $1000 per metric ton. In the middle of 2011, it was about $650 and later, the price of HFO further increased. Therefore, it can be said that the price of LNG is more stable compared to the HFO depending on the industry price [37].

Moreover, LNG is a very pure fuel; therefore, the operational costs of the engine are decreasing, the technical states of the engines are better, and the number of emergencies and failures are dropping. The price of LNG in the next 20 years will be equivalent to HFO, but later the LNG price will fall [37].

To make a proper comparison, different performance parameters of LNG-diesel dual-fuel engine such as torque, fuel consumption, smoke density, brake power and different emissions are needed to be measured and compared with a single diesel fuel engine (Table 4). For LNG-diesel dual-fuel engine, the brake specific fuel consumption (BSFC) is lower under light loads. Compared to the diesel engine, the torque and brake power of LNG-diesel dual-fuel engine remain unchanged [58]. Some researchers estimated brake thermal efficiency of the liquid natural gas engine has the same level of the corresponding conventional diesel engine [45]. Moreover, LNG is more preferable for heavy-duty vehicles compared to CNG as the density of LNG is 435 kg/m^3 as compared to 175 kg/m^3 for CNG at 20 MPa [58].

Table 4. Recent research on LNG as a fuel for the diesel engine.

Author	Working Environment	Outcome	Remarks
Kraipat Cheenkachorn et al. [3]	Used liquefied natural gas as primary fuel and small amount of diesel pilot was used as ignition source. Compared the result of dual fuel engine with equivalent diesel engine.	At 1100 rpm the torque obtained in duel fuel engine was around 1780 N-m and in diesel engine 1790 N-m. Additionally, the power output of both engines was equivalent. At lower rpm, thermal efficiency of both the engines is almost same.	LNG and diesel both showed similar power output.
Jiantong Song et al. [25]	Conducted research on LNG dual-fuel engine and conventional diesel engine. Compared the efficiency and emissions of both engines.	The brake specific fuel consumption of dual-fuel increases under <45.63 kW power. However, above this specific value, fuel consumption reduces. The smoke density of dual-fuel engine is significantly low compared to conventional diesel engine.	After a certain speed, the Brake specific fuel consumption (BSFC) of LNG engine reduces compared to diesel engine.
Ahmet Alper Yontar et al. [59]	Authors used Ricardo-Wave software for 1-D wide open throttle modeling of LNG and gasoline spark ignition engine for comparing performance and emissions.	Observed a torque value of 138 N-m and 110 N-m at 3000 and 3500 rpm for both gasoline and LNG respectively. At high speeds, the gasoline fuel consumption is comparatively higher than LNG consumption.	Both the fuels showed similar torque value.
Max Kofod et al. [60]	Conducted research on Well-to Wheel (WTT) Greenhouse Gas emissions of LNG used as a fuel for long haul trucks.	WTT GHG emission for LNG was 37.9 gCO_{2e}/MJ$_{out}$, whereas, for diesel was 47 gCO_{2e}/MJ$_{out}$. Total WTW GHG emission for LNG and diesel was 211.7 and 262 gCO_{2e}/MJ$_{out}$, respectively	GHG emissions of LNG fuel is lower than diesel fuel.
Junli Shi et al. [61]	Did a life cycle assessment to determine energy saving and environmental emission of a remanufactured LNG engine and newly manufactured diesel engine.	For a mileage of 300,000 km the fuel efficiency of diesel and LNG was 25 L/100 km and 26.5 m^3/100 km, respectively.	Both the systems showed almost similar fuel consumption.
Jeong Ok Han et al. [62]	The LNG engine (dual-fuel) was converted from a conventional diesel engine having 12 liter class and compared power, efficiency and emission	Compared to diesel engine, the power of LNG engine was 5% less, also the efficiency was lower. However, the LNG dual-fuel engine showed better results in terms of emissions.	Converted 12 liter class engines are not appropriate for using LNG.
Seokhwan Lee et al. [63]	The authors converted an electronically controlled diesel engine to dual-fuel engine and examined fuel economy, power etc.	Almost 85% of diesel substitution ratio was shown by the developed vehicle. The emission results were satisfactory and met k2006 standard. Moreover, the LNG dual-fuel engine performance was equivalent to the conventional diesel engine.	LNG showed same power output and less emissions compared to diesel.
Chandan Misra et al. [64]	Authors conducted research on two diesels and LNGs with three-way catalyst and one hydraulic hybrid diesel system and compared the NOx formation.	Found that the NOx emissions of LNG were slowest of all the technologies tested and emission from diesel was highest of all the technologies.	NOx emissions of LNG are lower than diesel.
Broynolf et al. [65]	Conducted research on the transportation of 1 t cargo 1 km with a ro-ro vessel and examined different parameters such as emissions, fuel characteristics etc.	CO_2 (fossil origin) (g/MJ fuel) for LNG was 8.3 and HFO was 6.7, CH_4 (g/MJ fuel) for HFO 0.072 and LNG 0.033. C_3H_8 (g/MJ fuel) for HFO 0.0067 and LNG 0.027. In case of emissions of air during tank-to-propeller-CO_2 (fossil origin; g/MJ fuel) for HFO 77 and LNG 54. NO_X (g/MJ fuel) for HFO 1.6 and LNG 0.11.	In terms of emissions, LNG shows better results (except fossil origin) than HFO.

Table 4. *Cont.*

Author	Working Environment	Outcome	Remarks
Jiehui Li et al. [66]	Designed a control system for LNG dual-fuel marine engine and compared different engine parameters with conventional diesel engine.	Calculated fuel flow rate from 600 to 1800 rpm and found that for the same speed, LNG consumption is lower than diesel consumption. Natural gas consumption rate increases with rpm, but decreases sharply once the speed overtakes 1400 rpm. Authors also found that cost of LNG dual-fuel engine is lower than diesel engine with a maximum decrease of 28.7% at 1300 rpm.	LNG engines show way better results than a conventional dual-fuel engine in terms of fuel consumption over 1400 rpm.
Ibrahim S. Seddiek et al. [67]	Conducted research on on-board diesel engine and natural gas (LNG) dual-fuel engine for ships.	Found that fuel consumption and all sorts of emissions (except hydrocarbon) are less for LNG dual-fuel compared to on-board diesel engine.	BSFC and emissions of LNG are less than diesel.
Green Truck Partnership project (report) [52]	Green Truck Partnership evaluates the quality and potential of fuels for heavy duty trucks. The discussed report analyzed LNG dual-fuel trial conducted under the program in 2013.	Throttle body injection LNG dual-fuel system did not produce any benefits regarding emissions. However, cost saving was around 4% compared to the conventional system. Authors recommended using the system in a case where the gas substitution rate could be higher.	Throttle body injection is not effective when using LNG where the gas substitution rate is low.
Harald Schlick [68]	Performed some experiments on LNG dual-fuel engine and compared power and NO_x emission.	Found that NOx and CO_2 emissions of LNG dual-fuel engines are clearly lower than diesel engine. They also recommended some solutions for reducing THC/CH_4/CO emissions: Valve overlap optimization, crevice volume reduction and minimization of flame quenching.	LNG shows better results in terms of emissions.
Dominik Schneiter et al. [69]	Performed an experiment on X-DF (engine model/series) LNG engine and compared with X-DF diesel engine.	GHG, NOx, SOx and particulate matter emission of X-DF LNG engine are less than diesel. SOx emission of LNG engine was found almost zero.	X-DF LNG engines show way better results than diesel in terms of emissions.
Hengbing Zhao et al. [70]	Analyzed emissions and power of class 8 hybrid electric truck technologies electricity, hydrogen, diesel and LNG as fuels for numerous applications.	At part load, the LNG compression ignition engine shows similar efficiency as diesel engine and at full load the efficiency is a bit low.	The efficiency of LNG and diesel are almost equivalent at part load.

2.3. Challenges

Though there are many merits of LNG, it also has some challenges to overcome. Frost burn occurs when LNG comes into contact with human skin because of being a cryogenic fluid. This phenomenon is also known as frostbite or cold burn [27,28]. Structural failure may occur due to metal cracks and metal embrittlement of the LNG storage tank, which rises by LNG spill [28]. Besides, different common usable materials such as plastic, carbon steel and rubber become brittle at an enormously cold temperature [27]. Since methane does not sustain breathing, thus, due to a large concentration of methane dispersed in air and vapor accumulated near the ground may rise to asphyxiation [28]. When LNG is released on water, it becomes less dense, which gives rise to rapid vaporization of liquefied natural gas and known as rapid phase transition (RPT) [37,71–73]. Consequently, due to the accidental release of LNG, a strong pressure wave is developed and causes heat radiation and burning clouds. Besides, higher flammability of LNG causes different fatal hazards [28]. LNG vaporizes readily when it is released from containment. Then, the liquid will be heated by the surroundings, thus, causing it to vaporize. The vapor will be mixed with the surrounding air and carried out downward causing a cloud. Eventually, it will be mixed with additional air and will be further diluted. The vapor cloud may ignite if the flammable portion comes in contact with a fire source. Secondary fires can be generated by the burn-backs of the primary vapor cloud and cause severe damage to the persons caught within the cloud [74].

According to a safety study regarding LNG, the maximum effective distance of LNG transportation with trucks is 230 m [28]. Flash fire of LNG causes truck tank rupture, which is related to the distance effect. Boiling liquid expanding vapor explosion (BLEVE) may occur with lethal effects during rupturing an LNG truck tank with elevated pressure [28]. Sometimes because of BLEVE, the released liquid immediately flashes and atomizes the resulting fireball. Though fireball lasts for only a couple of seconds, its effects can be very dangerous [74]. If LNG truck tank rupture occurs during engulfing fuel firing, then the maximum effect distance would be 190 m [28]. Pool fires occur due to immediate ignition of LNG, which can cause several damages to the surrounding equipment and burns to people caught within the cloud. The surface emissive power of LNG pool fires generally lies in the range of 220 ± 50 kW/m^2. However, a huge amount of smoke is produced during a large fire, which is very dangerous [27].

Due to having different density, LNG has the potential to layer in unstable strata within the tank. These strata have the potential of rolling over to stabilize the liquid in the tank. Due to normal heat leak, the longer LNG layer gets heated and changes density until it becomes lighter than the upper layer. As a result, a liquid rollover may occur with sudden vaporization of LNG. However, if the design of pressure relief systems is not adequate the excess pressure can result in cracks or other structural failures in the tank [71]. Since LNG is extremely cold, it can cause damage to eyes and tissue. Oxygen deficient hazard (ODH) results from the displace of air caused by the release of LNG [75]. Due to the accidental release of flammable liquid from pressurized containment, the leak takes the form of a spray of liquid droplets and vapors and if it is ignited the resulting fire is called torch fire. Torch fire possesses similar types of hazards as a pool fire. In some cases, for similar size pool and torch fire, the radiant heating power of a torch fire is frequently greater than that of a pool fire [74].

In a high pressure dual fuel engine, the combustion is nearly complete when LNG is injected at a pressure of 30–35 MPa with a small amount of diesel. However, when using in low pressure dual-fuel LNG is injected at a low pressure, which is comparable to the Otto cycle. A major disadvantage of the dual fuel diesel engine is the high amount of methane slip compared to the diesel engine. At high and medium loads the air fuel mixture enters the crevices and cylinder wall causes methane slip [76]. Moreover, at low speed the emission problem is more severe because of a low air-fuel ratio. This phenomenon therefore causes high methane slip and fuel consumption as a result of bulk quenching in the coldest areas of the combustion chamber [77]. In an Otto based dual fuel engine, at high power the methane slip is low, however, with the decrease of power, methane slip increases significantly [78].

3. Possible Solutions

Researchers all over the world are trying to find out some ways to face the challenges to use LNG and have already found some techniques as discussed below. The safety barriers, i.e., layers of protections, may have been used for preventing, controlling or mitigating undesired accidents or events. Similarly, for preventing the depth of catastrophic accident, different types of safeguards and protection layers have been used. Currently, the LNG industry uses several layers of protection for minimizing and controlling the consequences connected with vapor dispersions, LNG spills and subsequent fires and explosions [27]. To minimize cryogenic embrittlement, different types of special steels have been developed [28]. Moreover, currently, operators are using LNG rollover models, which can optimize boil-off costs by inducing density stratification and by doing so, converting a dangerous configuration into a potentially operational asset [71].

Basic responses to accidental LNG spills include detecting the spill, securing the origin and taking measures to prevent it from worsening. It is very important to secure the leak and the area, move the people away from the spill, prevent ignition sources and monitoring should be performed carefully until no vapor remains in the flammable limits [79]. A potential technique to reduce the size of the flammable vapor cloud is to increase the vapor dispersion generated by the liquefied gas spill [27]. The cloud generated by the spill of unconfined LNG on water travels at the speed of the wind before dispersion. Since it is denser than air, for land-based facilities, it has some advantages i.e., it can be easily controlled, though sometimes it can be a disadvantage if it takes longer to disperse [80]. Some experimental conditions of LNG spills on the water are summarized in Table 5. To determine low flammability limit (LFL), spill rate, volume, vaporization rate and atmospheric condition are taken into account [80]. However, reducing the pool surface area would be an effective solution to reduce LNG vaporization [27]. Water spray curtains may be used as a promising technique to mitigate many toxic and flammable LNG vapors by reducing the concentration of LNG vapor clouds [80]. A properly designed water curtain is capable of enhancing the dispersion of LNG vapor cloud and reducing the vapor cloud exclusion zone through mechanical effects, dilution and thermal effects. It is considered one of the most economical and efficient cloud control techniques [81]. However, the effectiveness of different water curtains is still widely unknown because of the temperature increase and LNG concentration reduction [80].

Table 5. LNG dispersion test on water [80].

Experiment	Spill Volume (m³)	Spill Rate (m³/min)	Pool Radius (m)	Downward Distance to LFL (m; maximum)
ESSO	0.73–10.2	18.9	7–14	442
Shell	27–193	2.7–19.3	NA	2250 (visual)
Maplin Sands	5–20	1.5–4	10	190 ± 20
Avocet (LLNL)	4.2–4.52	4	6.82–7.22	220
Burro (LLNL)	24–39	11.3–18.4	5	420
Coyote (LLNL)	8–28	14–19	Not reported	310
Falcon (LLNL)	20.6–66.4	8.7–30.3	Not reported	380

LLNL—Lawrence Livermore National Laboratory; note: reprinted with permission from the publisher (American Chemical Society—Energy and Fuels).

Three types of thermal coatings are used for exposure protection. For many years, as a thermal protective coating in refineries and petrochemical industries, concrete has been used [82]. To mitigate LNG vapor ignition and pool fire, high expansion foam application would be one of the effective solutions [82]. Additionally, to extinguish LNG fire dry chemicals such as sodium bicarbonate, potassium bicarbonate and urea potassium bicarbonate are a suitable accomplishment [83]. In most cases, for high expansion foam-controlled LNG pool fire dry chemical is useful, where the magnitude of the fire is small, the production of heat is reduced and fire fighters can apply dry chemical competently. To create greater firefighting capabilities, potassium bicarbonate is usually mixed with an aqueous film forming foam (AFFF) to create a dual agent system. However, inducing corrosions of exposed metals, low visibility after discharge, inducing breathing hazards, clogging ventilation filters, etc., are some limitations of dry chemicals [27].

A very effective method in managing LNG fires is the use of the medium (20:1 up to 200:1) and high (200:1 up to 1000:1) expansion foams. For LNG fires the effective ratio for high expansion foam should be at least 500:1. Gaz de France experience has demonstrated that the utilization of high expansion foam on LNG flames can diminish the fire stature by 60% and also the radiant heat can be decreased by around 90% [79]. Moreover, the results of a recent experiment show that exfoliated ZrP nanoplates can stabilize high expansion foam and this foam can increase the duration over which it needs to be replenished, while by exchanging heat with the vapor and foam and makes the vapor lighter, which ensures effective vapor dispersion [84]. Nevertheless, it is very important here to mention that applying water on LNG fire is not effective as it will not extinguish it; rather, the liquid will vaporize more quickly if the water is applied. Moreover, the fire will get hotter and burn faster if the water is applied to it [79]. Moreover, in recent years, some improvements have been done on LNG dual-fuel engine to improve efficiency, reduce pollution and improve sustainability as discussed in Table 6.

Table 6. Recent advancements on LNG dual-fuel engine.

Author	Working Environment	Outcome	Remarks
Koichi Watanabe [85]	A new engine was developed by Niigata, which is called dual fuel engine used two-types of fuel: gas and oil.	The NOx emission of the engine meets International Maritime Organization Tier II and Tier III requirements at diesel operation and gas operation, respectively.	The engine is capable of maintaining Tire II and Tire III emission standards.
GH Choi et al. [86]	Authors conducted some experiments on retrofitted LNG-diesel dual-fuel engine. The design of intake manifold was modified and electronic control system (ECU) was used to control amount of injected diesel fuel.	The modified system narrowed down the cylinder to cylinder variation by almost 60%.	The designed intake manifold was capable of reducing cylinder to cylinder variation.
Zheng Chen et al. [87]	Authors analyzed the effects of high compression ratio with hydrogen enrichment for the efficiency of LNG dual-fuel engine.	Reported that due to the effect of high compression ratio, cylinder pressure rises, ignition advances and shortens the combustion duration. Therefore, the process increases combustion stability and indicated efficiency.	Hydrogen enrichment increases the efficiency of LNG dual-fuel engine.
Jianqin Fu et al. [88]	Authors used a novel approach to improve the performance of LNG engine. LNG was purified into liquefied methane and then it was used as the engine fuel.	Ignition delay period is reduced and start of combustion advances. The torque of the engine increased by 9.5%, while the BSFC was reduced by almost 10.9%.	Since the octane number of methane is higher, the engine shows some better performance. However, since the pressure and temperature of the engine would also be high with high compression ratio, NOx emissions should be taken into consideration.
PA Davies et al. [89]	Authors developed a risk assessment model. The model was developed to determine the release likelihood and also provided guidance for the selection of appropriate safeguards and the prevention of leak.	Authors reported that calculating the likelihood of releases helps to identify where additional safeguards are necessary and would be effective.	The discussed risk assessment model should be implemented in LNG applications.
Qijun Tang et al. [90]	Author investigated the intake air supply system to improve accelerating and climbing performance. Authors designed a set of air supply systems and attached it with a LNG engine, then, its performance was tested.	Reported that at 1000 rpm, the torque was increased by 31% and specific gas consumption decreased by 1.64%. Additionally, the vehicle acceleration time was decreased by around 14.7–30%.	The intake air supply system is very effective for LNG engines.
Gyeung Ho Choi et al. [91]	Tested the performance and emissions of LNG dual-fuel engine with two different gas injectors. The main objective of the research was to gain economic benefits by replacing imported injector by local product.	Reported that the local product can operate satisfyingly with no knocking. The emission and engine performance were not compromised.	Local injectors are cost effective and provide quite similar output like imported ones.
Zunhua Zhang et al. [92]	Authors performed a numerical investigation on exhaust reforming characteristics of hydrogen production on the LNG marine engine.	For methane reforming reaction, inhibition of coke formation and hydrogen yield, higher mass ratio of water to fuel is advantageous. However, carbon monoxide is produced with higher exhaust gas recirculation process.	The developed numerical model is capable of discussing some fundamental parameters of LNG marine engines.
Chunhua Zhang et al. [93]	Authors investigated the effects of combustion duration characteristic on NOx production and brake thermal efficiency of diesel-LNG dual-fuel engine.	Authors reported that at low and medium speeds, the production of NOx is higher, whereas, when the centroid angle of combustion duration is before top dead centre, NOx emission is completely opposite.	In the LNG dual-fuel engine, the NOx emission is less when the centroid angle of combustion duration is before top dead centre.
Seokhwan Lee et al. [63]	Authors converted an electronically operated diesel engine into dual-fuel engine system. Maximum driving distance, fuel economy and emissions were examined. Authors also did an ND 13-mode test.	The engine meets k2006 regulations and the performance of the engine was similar to a conventional diesel engine.	Electronically operated diesel engine can be converted into dual-fuel engine system without having any performance reduction.

Table 6. *Cont.*

Author	Working Environment	Outcome	Remarks
Sinian He et al. [94]	The authors proposed and investigated a combined organic Rankine cycle (ORC) system, where exhaust waste was used as a heat source and LNG as a heat sink to provide alternative power for a LNG fired vehicle. In this study, five types of organic fluids were analyzed such as CF3I, R236EA, R236FA, RC318 and C4F10.	Reported that fluid R236FA provides the highest thermal efficiency (21.6%). These five fluids can improve fuel economy by more than 14.7%.	The study reported five types of fluids, which can increase combined organic Rankine cycle efficiency.
Yifeng Guan et al. [95]	Authors developed a fault tree model to analyze fire and explosions in a dual-fuel ship.	According to the faults found, authors suggested ten fundamental safety measures where authors put great importance on the working humans near accidents.	The study suggested some techniques, which could reduce the tendency of accidents.
Elena Stefana et al. [96]	Authors did a qualitative risk assessment on LNG-diesel dual-fuel engine. First authors developed a reliability block diagram, then performed failure mode analysis, failure effect analysis, likelihood and consequence analysis, fault tree and bow-tie analysis.	Bow tie analysis allowed providing barriers to prevent and mitigate critical events. By applying all the methods, authors were able to identify and design a set of safety measures.	Authors successfully developed some safety measures for the failures of LNG-diesel dual-fuel engine.
Khaled Senary et al. [97]	Authors developed a waste heat recovery system to meet IMO (international maritime organization) regulations onboard LNG carrier.	The developed system meets the requirements and regulations set by IMO for Tier-III. The waste heat recovery system is capable of reducing almost 130 kg NOx per day.	The waste heat recovery system is capable of reducing a huge amount of NOx production.

3.1. Engine Knock Reduction

Knocking is one of the severe problems of the LNG dual-fuel engine. One of the key initiatives to utilize LNG as a fuel in a diesel engine is to alleviate the knocking problem and take adequate precautions when knocking resistance is low. However, there is no specific technique to reduce knocking completely. One of the solutions to improve the knock resistance of LNG is to keep higher methane in LNG. During the manufacture of LNG, it should be ensured that the amount of methane is 99.4% or higher [14]. The knock resistance of LNG is strongly dependent upon the boil-off rate of the cargo container system used and the initial composition of the fuel. The reduction of methane and nitrogen in LNG increases the fraction of butane and propane and results in decrease in the PKI-MN (propane knock index-methane number), which is responsible for knocking [23]. Further, to reduce the knock combustion, leaning combustible mixture and exhaust gas recirculation strategy can be applied as an efficient way [98]. In the case of an SI engine, higher turbulent kinetic energy in the centre of the combustion chamber, lower mean flow velocity in the spark plug region and near the spark plug can produce knock [99]. Therefore, it is necessary to develop combustion chamber, which can provide higher mean flow velocity.

3.2. Future Research Perspective

With the increasing demand and potential of LNG, numerous research has been conducted regarding the use of it in a diesel engine. According to previous discussion and literature, researchers and engineers have successfully used LNG in a diesel engine, where the emissions are less, and the power output is similar to conventional engines. However, more research is necessary regarding fuel consumption, fuel atomization, engine equipment and safety of using LNG. According to research [19], below the power of 45.63 kW, the BSFC of LNG dual-fuel engine is higher, though after this certain limit BSFC tends to reduce. The reason can be that the LNG–air mixture may be too lean to burn under light loads. However, as the power increases, the mixture of fuel in the cylinder enriches, eventually the conditions of combustion are improved, which reduces BSFC. Further research is necessary on how the BSFC of LNG dual-fuel engine can be reduced under light loads. Literature suggest that under a certain load and power output conditions, formation of hydrocarbon is almost double in LNG-dual fuel engine than a conventional on board diesel engine, which also should be a concern for the researchers. Research related to different powertrain such as series and parallel hybrids associated with the use of LNG are very rare. Hengbing Zhao et al. [99] did research, where the authors developed and analyzed some hybrid configurations and according to the simulation results, LNG-CI hybrids showed better results in terms of emissions for Class 8 Hybrid-Electric Truck. Experiments should be conducted in this area to verify the results. Moreover, the hazards and dynamics of LNG spill require more research. Regarding vaporizers, the IFV shows various advantages over other types. In the above discussed literature, finding related to design, configurational variations, working fluids, etc., are discussed. However, it is recommended to develop a three-dimensional unsteady CFD model to evaluate instantaneous flow and heat transfer. As discussed above, in the LNG dual-fuel engine, at low and medium speeds, the production of NOx is higher. Advancing the diesel injection timing, increases the proportion of compression and eventually increases the peak in-cylinder temperature. At medium loads with the increase of temperature, energy in the exhaust gas increases, which eventually increases the intake pressure. This also should be a concern of researchers and further research is necessary to develop a solution to mitigate this problem. For determining hazards, experimental data from large scale experiments are not sufficient. Therefore, it is recommended to consider small scale experiments where large scale is not feasible.

4. Conclusions

All over the world, liquefied natural gas (LNG) would be a clean primary energy source shortly. LNG provides a plethora of potential point of interests as a fuel for a dual fuel engine. It is capable of compensating some of the major drawbacks of natural gas and diesel vehicles. LNG is not only clean, has relatively higher density, has easier transportation and is safe but also the price is reasonable. The indicated thermal efficiency and brake thermal efficiency of an engine using LNG are equivalent to that of diesel and the production of emission is less when using LNG. This paper also reviews successfully the challenges, finds out some responses to overcome those challenges and provides some recommendations for proper utilization of LNG as a fuel for the dual fuel engine. The major challenges of LNG are its higher flammability causes different fatal hazards and when using in dual-fuel engine causes knock. Though researchers have been successful to find out some ways to overcome some challenges, further research is necessary to reduce the hazards and make the fuel more effective and environment-friendly when using as a fuel for a diesel engine.

Author Contributions: Conceptualization, M.A.A. and M.N.N.; formal analysis, M.A.A. and M.T.I.; investigation, M.A.A. and M.N.N.; resources, M.A.A. M.W.A.; data curation, M.A.A., M.T.I. and M.W.C.; writing—original draft preparation, M.A.A. and M.N.N.; writing—review & Editing- M.A.A., M.N.N., M.T.I., M.W.C. and M.W.A.; visualization, M.A.A., M.N.N., M.T.I., M.W.C. and M.W.A.; project administration, M.A.A. and M.N.N.; All authors have read and agreed to the published version of the manuscript.

Funding: This research received no external funding

Conflicts of Interest: The authors declare no conflict of interest

References

1. Rafiee, A.; Khalilpour, K.R. Renewable Hybridization of Oil and Gas Supply Chains. In *Polygeneration with Polystorage for Chemical and Energy Hubs*; Academic Press: Cambridge, MA, USA, 2019; pp. 331–372. [CrossRef]
2. Nabi, M.N.; Hustad, J.E. Influence of Biodiesel Addition to Fischer—Tropsch Fuel on Diesel Engine Performance and Exhaust Emissions. *Energy Fuels* **2010**, *24*, 2868–2874. [CrossRef]
3. Cheenkachorn, K.; Poompipatpong, C.; Ho, C.G. Performance and emissions of a heavy-duty diesel engine fuelled with diesel and LNG (liquid natural gas). *Energy* **2013**, *53*, 52–57. [CrossRef]
4. Nabi, M.N.; Hustad, J.E. Experimental investigation of engine emissions with marine gas oil-oxygenate blends. *Sci. Total Environ.* **2010**, *408*, 3231–3239. [CrossRef]

5. Zare, A.; Bodisco, T.A.; Nabi, M.N.; Hossain, F.M.; Ristovski, Z.D.; Brown, R.J. A comparative investigation into cold-start and hot-start operation of diesel engine performance with oxygenated fuels during transient and steady-state operation. *Fuel* **2018**, *228*, 390–404. [CrossRef]

6. Djermouni, M.; Ouadha, A. Comparative assessment of LNG and LPG in HCCI engines. *Energy Procedia* **2017**, *139*, 254–259. [CrossRef]

7. Kumar, S.; Kwon, H.-T.; Choi, K.-H.; Lim, W.; Cho, J.H.; Tak, K.; Moon, I. LNG: An eco-friendly cryogenic fuel for sustainable development. *Appl. Energy* **2011**, *88*, 4264–4273. [CrossRef]

8. Pfoser, S.; Aschauer, G.; Simmer, L.; Schauer, O. Facilitating the implementation of LNG as an alternative fuel technology in landlocked Europe: A study from Austria. *Res. Transp. Bus. Manag.* **2016**, *18*, 77–84. [CrossRef]

9. Verbeek, R.; Kadijk, G.; Van Mensch, P.; Wulffers, C.; Van den Beemt, B.; Fraga, F.; Aalbers, A.D.A. *Environmental and Economic Aspects of Using LNG as a Fuel for Shipping in The Netherlands*; TNO: Delft, The Netherlands, 2011.

10. Henderson, J. *The Prospects for Future LNG Supply Outside Australia and the USA*; Oxford Energy Forum; Oxford Institute for Energy Studies: Oxford, UK, 2017.

11. Dutta, A.; Karimi, I.A.; Farooq, S. Economic Feasibility of Power Generation by Recovering Cold Energy during LNG (Liquefied Natural Gas) Regasification. *ACS Sustain. Chem. Eng.* **2018**, *6*, 10687–10695. [CrossRef]

12. Winter, J.; Dobson, S.; Fellows, G.K.; Lam, D.; Craig, P. An Overview of Global Liquefied Natural Gas Markets and Implications for Canada. *School Public Policy Publ.* **2018**, *11*, 1–27.

13. *World LNG Estimated Landed Price: June 19*; Federal Energy Regulatory Commission-Waterborne Energy, Waterborne Energy: Houston, TX, USA, 2019.

14. Frailey, M. *Using LNG as a Fuel in Heavy-Duty Tractors*; NREL/SR-540-24146; NREL: Golden, CO, USA, 1999.

15. Arteconi, A.; Brandoni, C.; Evangelista, D.; Polonara, F. Life-cycle greenhouse gas analysis of LNG as a heavy vehicle fuel in Europe. *Appl. Energy* **2010**, *87*, 2005–2013. [CrossRef]

16. Mansour, C.; Bounif, A.; Aris, A.; Gaillard, F. Gas—Diesel (dual-fuel) modeling in diesel engine environment. *Int. J. Therm. Sci.* **2001**, *40*, 409–424. [CrossRef]

17. Nwafor, O.M. Knock characteristics of dual-fuel combustion in diesel engines using natural gas as primary fuel. *Sadhana* **2002**, *27*, 375–382. [CrossRef]

18. Nwafor, O.M. Effect of advanced injection timing on emission characteristics of diesel engine running on natural gas. *Renew. Energy* **2007**, *32*, 2361–2368. [CrossRef]

19. Karim, G.A. Combustion in Gas-fueled Compression Ignition Engines of the Dual Fuel Type. *Handb. Combust* **2010**, 213–235.

20. Abd Alla, G.H.; Soliman, H.A.; Badr, O.A.; Abd Rabbo, M.F. Effect of injection timing on the performance of a dual fuel engine. *Energy Conversat. Manag.* **2002**, *43*, 269–277. [CrossRef]

21. Selim, M.Y.E. Sensitivity of dual fuel engine combustion and knocking limits to gaseous fuel composition. *Energy Convers. Manag.* **2004**, *45*, 411–425. [CrossRef]

22. Selim, M.Y.E. Pressure—Time characteristics in diesel engine fueled with natural gas. *Renew. Energy* **2001**, *22*, 473–489. [CrossRef]

23. Van Essen, M.; Gersen, S.; Van Dijk, G.; Levinsky, H.; Mundt, T.; Dimopoulos, G.; Kakalis, N. *The Effect of Boil off on the Knock Resistance of LNG Gases*; CIMAC Congress: Helsinki, Finland, 2016; pp. 123–127.

24. Verbeek, R.; Verbeek, M. *LNG for Trucks and Ships: Fact Analysis Review of Pollutant and GHG Emissions Final*; TNO Innov. Life: The Hague, The Netherlands, 2015.

25. Song, J.; Yao, J.; Lv, J. Performance and an optimisation control scheme of a heavy-duty diesel engine fuelled with LNG-diesel dual-fuel. *Int. J. Heavy Veh. Syst.* **2018**, *25*, 189–202.

26. Mokhatab, S.; Mak, J.Y.; Valappil, J.V.; Wood, D.A. *Handbook of Liquefied Natural Gas*; Gulf Professional Publishing: Houston, TX, USA, 2013.

27. Mannan, S. Liquefied Natural Gas. In *Lees' Process Safety Essentials*; Gulf Professional Publishing: Oxford, UK, 2014; pp. 431–435.

28. Vandebroek, L.; Berghmans, J. Safety Aspects of the use of LNG for Marine Propulsion. *Procedia Eng.* **2012**, *45*, 21–26. [CrossRef]

29. Kumar, S.; Kwon, H.-T.; Choi, K.-H.; Hyun Cho, J.; Lim, W.; Moon, I. Current status and future projections of LNG demand and supplies: A global prospective. *Energy Policy* **2011**, *39*, 4097–4104. [CrossRef]

30. *Basic Properties of LNG*; LNG Information Paper No-1; GIIGNL: Levallois, France, 2019.
31. Rathnayaka, S.; Khan, F.; Amyotte, P. Accident modeling approach for safety assessment in an LNG processing facility. *J. Loss Prev. Process Ind.* **2012**, *25*, 414–423. [CrossRef]
32. Bahadori, A. Liquefied Natural Gas (LNG). In *Natural Gas Processing*; Elsevier: Amsterdam, The Netherlands, 2014; pp. 591–632.
33. *Liquefied Natural Gas Fuel System Users' Manual*; Agility Fuel Solutions: Santa Ana, CA, USA, 2017.
34. Shi, G.-H.; Jing, Y.-Y.; Wang, S.-L.; Zhang, X.-T. Development status of liquefied natural gas industry in China. *Energy Policy* **2010**, *38*, 7457–7465. [CrossRef]
35. Lin, W.; Zhang, N.; Gu, A. LNG (liquefied natural gas): A necessary part in China's future energy infrastructure. *Energy* **2010**, *35*, 4383–4391. [CrossRef]
36. Won, W.; Lee, S.K.; Choi, K.; Kwon, Y. Current trends for the floating liquefied natural gas (FLNG) technologies. *Korean J. Chem. Eng.* **2014**, *31*, 732–743. [CrossRef]
37. Herdzik, J. LNG as a marine fuel-possibilities and problem. *J. KONES* **2011**, *18*, 169–176.
38. Alternative Fuels Data Center—Fuel Properties Comparison. 2014. Available online: https://afdc.energy.gov/fuels/fuel_comparison_chart.pdf (accessed on 21 September 2019).
39. Laugen, L. An Environmental Life Cycle Assessment of LNG and HFO as Marine Fuels. Master's Thesis, Institutt for Marin Teknikk, Norwegian University of Science & Technology, Trondheim, Norway, 2013.
40. Osorio-Tejada, J.L.; Llera-Sastresa, E.; Scarpellini, S. Liquefied natural gas: Could it be a reliable option for road freight transport in the EU? *Renew. Sustain. Energy Rev.* **2017**, *71*, 785–795. [CrossRef]
41. Hao, H.; Liu, Z.; Zhao, F.; Li, W. Natural gas as vehicle fuel in China: A review. *Renew. Sustain. Energy Rev.* **2016**, *62*, 521–533. [CrossRef]
42. Kanbur, B.B.; Xiang, L.; Dubey, S.; Choo, F.H.; Duan, F. Cold utilization systems of LNG: A review. *Renew. Sustain. Energy Rev.* **2017**, *79*, 1171–1188. [CrossRef]
43. Strantzali, E.; Aravossis, K.; Livanos, G.A. Evaluation of future sustainable electricity generation alternatives: The case of a Greek island. *Renew. Sustain. Energy Rev.* **2017**, *76*, 775–787. [CrossRef]
44. Bakas, I. *Propulsion and Power Generation of LNG driven Vessels*; University of Piraeus: Pireas, Greece, 2015.
45. Goto, Y. Development of a liquid natural gas pump and its application to direct injection liquid natural gas engines. *Int. J. Engine Res.* **2002**, *3*, 61–68. [CrossRef]
46. Khan, M.I.; Yasmin, T.; Shakoor, A. Technical overview of compressed natural gas (CNG) as a transportation fuel. *Renew. Sustain. Energy Rev.* **2015**, *51*, 785–797. [CrossRef]
47. Sonthalia, A.; Rameshkumar, C.; Sharma, U.; Punganur, A.; Abbas, S. Combustion and performance characteristics of a small spark ignition engine fuelled with HCNG. *J. Eng. Sci. Technol.* **2015**, *10*, 404–419.
48. Lemaire, R.; Faccinetto, A.; Therssen, E.; Ziskind, M.; Focsa, C.; Desgroux, P. Experimental comparison of soot formation in turbulent flames of Diesel and surrogate Diesel fuels. *Proc. Combust. Inst.* **2009**, *32*, 737–744. [CrossRef]
49. Hanshaw, G.; Pope, G. *Liquefied Natural Gas Criteria/Comparative Values for Use as an Automotive Fuel*; SAE Technical Paper 0148-7191; SAE: Warrendale, PA, USA, 1996.
50. Jeff, C. *Natural Gas Fuel System LNG. Product Information Bulletins*; Cummins Westport: Vancouver, BC, Canada, 2012.
51. Hernes, H.E. Active and Passive Measures to Maintain Pressure in LNG Fuel Systems for Ships. Master's Thesis, NTNU, Trondheim, Norway, 2015.
52. Xu, S.; Chen, X.; Fan, Z. Design of an Intermediate Fluid Vaporizer for Liquefied Natural Gas. *Chemical Eng. Technol.* **2017**, *40*, 428–438. [CrossRef]
53. Bassi, A. *Liquefied Natural Gas (LNG) as Fuel for Road Heavy Duty Vehicles Technologies and Standardization*; SAE Technical Paper Series; SAE: Warrendale, PA, USA, 2011.
54. Burel, F.; Taccani, R.; Zuliani, N. Improving sustainability of maritime transport through utilization of Liquefied Natural Gas (LNG) for propulsion. *Energy* **2013**, *57*, 412–420. [CrossRef]
55. Brett, B.C. *Potential Market for LNG-Fueled Marine Vessels in the United States*; Massachusetts Institute of Technology: Cambridge, MA, USA, 2008.
56. Rosenstiel, D.V.; Bünger, U.; Schmidt, P.R.; Weindorf, W.; Wurster, R.; Zerhusen, J. *LNG in Germany: Liquefied Natural Gas and Renewable Methane in Heavy-Duty Road Transport*; German Energy Agency: Berlin, Germany, 2014.

57. Cluzel, C.; Riley, R. *Development of a Well to Tank Emission Tool for Heavy Goods Vehicles*; Element Energy Ltd.: Cambridge, UK, 2018.

58. Song, J.; Zhang, C.; Lin, G.; Zhang, Q. Performance and emissions of an electronic control common-rail diesel engine fuelled with liquefied natural gas-diesel dual-fuel under an optimization control scheme. *Proc. Inst. Mech. Eng. Part D J. Automob. Eng.* **2018**, *233*, 1380–1390. [CrossRef]

59. Yontar, A.A.; Doğu, Y. 1-D modelling comparative study to evaluate performance and emissions of a spark ignition engine fuelled with gasoline and LNG. In *MATEC Web of Conferences*; EDP Sciences: Les Ulis, France, 2016; p. 05003.

60. Kofod, M.; Stephenson, T. *Well-to Wheel Greenhouse Gas Emissions of LNG Used as a Fuel for Long Haul Trucks in a European Scenario*; SAE Technical Paper 0148-7191; SAE: Warrendale, PA, USA, 2013.

61. Shi, J.; Li, T.; Liu, Z.; Zhang, H.; Peng, S.; Jiang, Q.; Yin, J. Life cycle environmental impact evaluation of newly manufactured diesel engine and remanufactured LNG engine. *Procedia CIRP* **2015**, *29*, 402–407. [CrossRef]

62. Han, J.-O.; Chae, J.-M.; Lee, J.-S.; Hong, S.-H. Economical Evaluation of a LNG Dual Fuel Vehicle Converted from 12L Class Diesel Engine. *J. Energy Eng.* **2010**, *19*, 246–250.

63. Lee, S.H.; Lee, J.W.; Heo, S.J.; Yoon, S.S.; Roh, Y.H. Characteristics of Electronically Controlled 13L LNG-Diesel Dual Fuel Engine. *J. Korean Inst. Gas* **2007**, *11*, 54–58.

64. Misra, C.; Ruehl, C.; Collins, J.; Chernich, D.; Herner, J. In-use NOx emissions from diesel and liquefied natural gas refuse trucks equipped with SCR and TWC, respectively. *Environ. Sci. Technol.* **2017**, *51*, 6981–6989. [CrossRef] [PubMed]

65. Brynolf, S.; Fridell, E.; Andersson, K. Environmental assessment of marine fuels: Liquefied natural gas, liquefied biogas, methanol and bio-methanol. *J. Clean. Prod.* **2014**, *74*, 86–95. [CrossRef]

66. Li, J.; Wu, B.; Mao, G. Research on the performance and emission characteristics of the LNG-diesel marine engine. *J. Nat. Gas Sci. Eng.* **2015**, *27*, 945–954. [CrossRef]

67. Seddiek, I.S.; Elgohary, M.M. Eco-friendly selection of ship emissions reduction strategies with emphasis on SOx and NOx emissions. *Int. J. Nav. Archit. Ocean Eng.* **2014**, *6*, 737–748. [CrossRef]

68. Schlick, H. Potentials and challenges of gas and dual-fuel engines for marine application. In Proceedings of the 5th CIMAC Cascades, Busan, Korea, 23 October 2014; pp. 1–31.

69. Schneiter, D.; Nylund, I. *Greenhouse Gas (GHG) Emissions from LNG Engines. Review of the Two Stroke Engine Emission Footprint. 4—Emission Reduction Technologies—What's in Store for the Future*; Paper 426; Cimac Congress: Vancouver, Canada, 2019.

70. Zhao, H.; Burke, A.; Zhu, L. Analysis of Class 8 hybrid-electric truck technologies using diesel, LNG, electricity, and hydrogen, as the fuel for various applications. In *2013 World Electric Vehicle Symposium and Exhibition (EVS27)*; IEEE: Barcelona, Spain, 2013; pp. 1–16.

71. Alderman, J.A. Introduction to LNG safety. *Process Saf. Prog.* **2005**, *24*, 144–151. [CrossRef]

72. Qiao, Y.; West, H.H.; Mannan, M.S.; Johnson, D.W.; Cornwell, J.B. Assessment of the effects of release variables on the consequences of LNG spillage onto water using FERC models. *J. Hazard. Mater.* **2006**, *130*, 155–162. [CrossRef]

73. Rana, M.A. Forced Dispersion of Liquefied Natural Gas Vapor Clouds with Water Spray Curtain Application. Ph.D. Thesis, Texas A&M University, College Station, TX, USA, 2009.

74. Zinn, C.D. LNG codes and process safety. *Process Saf. Prog.* **2005**, *24*, 158–167. [CrossRef]

75. Peterson, T.J.; Weisend, J., II. Liquefied Natural Gas (LNG) Safety. In *Cryogenic Safety*; Springer: Berlin/Heidelberg, Germany, 2019; pp. 181–189.

76. Krivopolianskii, V.; Valberg, I.; Stenersen, D.; Ushakov, S.; Æsøy, V. Technology, Control of the Combustion Process and Emission Formation in Marine Gas Engines. *J. Mar. Sci. Technol.* **2019**, *24*, 593–611. [CrossRef]

77. Lindstad, E.; Eskeland, G.S.; Rialland, A.; Valland, A. Decarbonizing Maritime Transport: The Importance of Engine Technology and Regulations for LNG to Serve as a Transition Fuel. *Sustainability* **2020**, *12*, 8793. [CrossRef]

78. Ushakov, S.; Stenersen, D.; Einang, P.M. Methane slip from gas fuelled ships: A comprehensive summary based on measurement data. *J. Mar. Sci. Technol.* **2019**, *24*, 1308–1325. [CrossRef]

79. Walker, A.H. *Response Considerations for LNG Spills*; Interspill Conference: London, UK, 2006.

80. Ikealumba, W.C.; Wu, H. Some Recent Advances in Liquefied Natural Gas (LNG) Production, Spill, Dispersion, and Safety. *Energy Fuels* **2014**, *28*, 3556–3586. [CrossRef]

81. Rana, M.A.; Guo, Y.; Mannan, M.S. Use of water spray curtain to disperse LNG vapor clouds. *J. Loss Prev. Process. Ind.* **2010**, *23*, 77–88. [CrossRef]

82. Lees, F. *Lees' Loss Prevention in the Process Industries: Hazard Identification, Assessment and Control*; Butterworth-Heinemann, Elsevier: Amsterdam, The Netherlands, 2012.

83. Dalaklis, D. Effective fire-fighting strategies for LNG during bunkering. *World Marit. Univ.* Available online: https://www.onthemosway.eu/wp-content/uploads/2015/09/PRESENTATION-1-%E2%80%93-EFFECTIVE-FIRE-FIGHTING-STRATEGIES-FOR-LNG-DURING-BUNKERING.pdf (accessed on 1 November 2020).

84. Krishnan, P.; Al-Rabbat, A.; Zhang, B.; Huang, D.; Zhang, L.; Zeng, M.; Mannan, M.S.; Cheng, Z.J.P.S.; Protection, E. Improving the stability of high expansion foam used for LNG vapor risk mitigation using exfoliated zirconium phosphate nanoplates. *Process. Saf. Environ. Prot.* **2019**, *123*, 48–58. [CrossRef]

85. Watanabe, K. High operation capable marine dual fuel engine with LNG. *Mar. Eng.* **2015**, *50*, 738–743. [CrossRef]

86. Choi, G.; Hwang, S.; Poompipatpong, C.; Lee, S.; Kim, E. A Study on Cylinder-to-Cylinder Variations in Retrofitted LNG-Diesel Dual Fueled Engine. In Proceedings of the 1st International Conference of Multi Disciplines of Engineering on Advanced Technology and Environmentalism Design, Dusit Thani Hotel Pattaya, Pattaya, Thailand, 6–8 March 2012.

87. Chen, Z.; Xu, B.; Zhang, F.; Liu, J. Quantitative research on thermodynamic process and efficiency of a LNG heavy-duty engine with high compression ratio and hydrogen enrichment. *Appl. Therm. Eng.* **2017**, *125*, 1103–1113. [CrossRef]

88. Fu, J.; Shu, J.; Zhou, F.; Liu, J.; Xu, Z.; Zeng, D. Experimental investigation on the effects of compression ratio on in-cylinder combustion process and performance improvement of liquefied methane engine. *Appl. Therm. Eng.* **2017**, *113*, 1208–1218. [CrossRef]

89. Davies, P.; Fort, E. Technology, LNG as a marine fuel: Likelihood of LNG releases. *J. Mar. Eng. Technol.* **2013**, *12*, 3–10.

90. Tang, Q.; Fu, J.; Liu, J.; Zhou, F.; Yuan, Z.; Xu, Z. Performance improvement of liquefied natural gas (LNG) engine through intake air supply. *Appl. Therm. Eng.* **2016**, *103*, 1351–1361. [CrossRef]

91. Choi, G.H.; Tangsiriworakul, C.; Poompipatpong, C. Performance and exhaust emission studies of a large LNG-diesel engine operating with different gas injector's characteristics. *KMUTNB Int. J. Appl. Sci. Technol.* **2014**, *7*, 59–66.

92. Zhang, Z.; Jia, P.; Zhong, G.; Liang, J.; Li, G. Numerical study of exhaust reforming characteristics on hydrogen production for a marine engine fueled with LNG. *Appl. Therm. Eng.* **2017**, *124*, 241–249. [CrossRef]

93. Zhang, C.; Zhou, A.; Shen, Y.; Li, Y.; Shi, Q. Effects of combustion duration characteristic on the brake thermal efficiency and NOx emission of a turbocharged diesel engine fueled with diesel-LNG dual-fuel. *Appl. Therm. Eng.* **2017**, *127*, 312–318. [CrossRef]

94. He, S.; Chang, H.; Zhang, X.; Shu, S.; Duan, C. Working fluid selection for an Organic Rankine Cycle utilizing high and low temperature energy of an LNG engine. *Appl. Therm. Eng.* **2015**, *90*, 579–589. [CrossRef]

95. Guan, Y.; Zhao, J.; Shi, T.; Zhu, P. Fault tree analysis of fire and explosion accidents for dual fuel (diesel/natural gas) ship engine rooms. *J. Mar. Sci. Appl.* **2016**, *15*, 331–335. [CrossRef]

96. Stefana, E.; Marciano, F.; Alberti, M. Qualitative risk assessment of a Dual Fuel (LNG-Diesel) system for heavy-duty trucks. *J. Mar. Sci. Appl.* **2016**, *39*, 39–58. [CrossRef]

97. Senary, K.; Tawfik, A.; Hegazy, E.; Ali, A. Development of a waste heat recovery system onboard LNG carrier to meet IMO regulations. *J. Loss Prev. Process. Ind.* **2016**, *55*, 1951–1960. [CrossRef]

98. Bereczky, Á. *Investigation of Knock Limits of Dual Fuel Engine*; Department of Energy, Budapest University of Technology and Economics (BME): Budapest, Hungary, 2013.

99. Zhao, X.; Wang, H.; Zheng, Z.; Yao, M.; Sheng, L.; Zhu, Z. *Evaluation of Knock Intensity and Knock-Limited Thermal Efficiency of Different Combustion Chambers in Stoichiometric Operation LNG Engine*; SAE Technical Paper 0148-7191; SAE: Warrendale, PA, USA, 2019.

Publisher's Note: MDPI stays neutral with regard to jurisdictional claims in published maps and institutional affiliations.

MDPI

St. Alban-Anlage 66

4052 Basel

Switzerland

Tel. +41 61 683 77 34

Fax +41 61 302 89 18

www.mdpi.com

Energies Editorial Office

E-mail: energies@mdpi.com

www.mdpi.com/journal/energies

www.ingramcontent.com/pod-product-compliance
Lightning Source LLC
LaVergne TN
LVHW080501180726
843515LV00016B/592